Engineering
Problem-Solv

101

It is difficult to say what is impossible,
for the dream of yesterday is the hope of today and reality of tomorrow.
ROBERT GODDARD (1882–1945),
first to create and launch a liquid-fueled rocket on March 16, 1926

Engineering Problem-Solving 101

Time-Tested and Timeless Techniques

Robert W. Messler, Jr.

New York   Chicago   San Francisco   Lisbon   London   Madrid
Mexico City   Milan   New Delhi   San Juan   Seoul
Singapore   Sydney   Toronto

The McGraw·Hill Companies

1 2 3 4 5 6 7 8 9 0   QVR/QVR   1 8 7 6 5 4 3 2

ISBN 978-0-07-179996-6
MHID 0-07-179996-6

Sponsoring Editor Michael Penn	**Copy Editor** Virginia E. Carroll, North Market Street Graphics
Editing Supervisor Stephen M. Smith	**Proofreaders** Stewart Smith and Virginia Landis, North Market Street Graphics
Production Supervisor Pamela A. Pelton	**Art Director, Cover** Jeff Weeks
Acquisitions Coordinator Bridget L. Thoreson	**Composition** North Market Street Graphics
Project Manager Virginia E. Carroll, North Market Street Graphics	

Printed and bound by Quad/Graphics.

To Asher B
Know that the engineer who chose to teach will love you forever . . .
Bop Bop

ABOUT THE AUTHOR

Robert W. Messler, Jr., Ph.D., is a recipient of numerous awards for teaching materials, welding, joining, and design at Rensselaer Polytechnic Institute. He practiced engineering for 16 years in industry, and then for 7 years he served as Technical Director of RPI's world-renowned Center for Manufacturing Productivity. Dr. Messler has authored five other technical books and is a fellow of both ASM International and the American Welding Society.

Contents

Preface

INHABITANTS OF OUR planet have been faced with problems that need to be overcome since they first walked erect . . . and, perhaps, before. Even many of our fellow creatures are faced with problems that have to be overcome, from building a nest, den, or lair, to finding and securing food, to fording streams, crossing swampy ground, or surviving storms. But human beings have distinguished themselves as problem solvers from time immemorial. From the "architects" (actually, engineers) who designed and built the Great Pyramids at Giza to the myriad of engineers of many disciplines who allowed a man to plant an American flag on the Moon before returning safely to Earth, mind-boggling problems—that most of these people would have referred to as "challenges," as opposed to problems, in their optimistic outlook—had to be overcome by a combination of ingenuity, creativity, agility, and versatility.

While a formal education in the sciences, mathematics, and basics of engineering (statics, dynamics, fluid mechanics, thermodynamics, kinetics, heat transfer, electronics, etc.), followed by discipline-specific required and elective courses, is necessary for a young person to become an engineer, this is not sufficient. To become a *real* engineer, one needs to "marinate" in engineering by working with and among other engineers, technicians, and skilled crafts- and tradespersons, learning the *techniques* of problem-solving that were probably not explicitly taught, if they were even mentioned, in engineering school. It is knowledge of these techniques that allows one to both solve a problem and, in the longer haul, approach problem-solving with confidence. Knowing *what to do* to solve a problem is empowering, while solving the problem is often only momentarily fulfilling.

It is the premise of this book, which the author derived from more than 25 years as an "engineer who teaches," that simply being given innumerable problems within courses to solve in a variety of subject areas may not enlighten the student with recognition of the *technique* that was employed to allow success. After all, it is a stark reality that an engineering student's first objective is to survive engineering school. His or her second objective—despite what professional educators and parents, alike, desire—is learning for the long haul. Even more significant may be that faculty in modern engineering schools or colleges build careers on the success of their research and fulfill an essential need by teaching what they know best. What they know best is, of course, their area of specialization, on which their research is based. Fewer and fewer of those who enter an academic career in engineering do so from a former position as a practicing engineer. Rather, more and more, they do so either directly from graduate school or through a carefully structured set of experiences as researchers (e.g., graduate research assistants, postdocs, tenure-track/tenured faculty). Thus, the reality is, fewer and fewer of those charged with teaching engineering actually practiced engineering; that is, they never "marinated" in engineering. Not surprisingly, therefore, fewer and fewer are explicitly aware of the variety of techniques by which real-world problems can be solved—beyond mathematical approaches based largely on "plug-and-chug" manipulations of equations using given values to arrive at a closed-ended solution.

This book is intended to end this situation—to attempt to shortcut the process of gaining the self-confidence that comes with having a diverse set of powerful tools and techniques by which problems can be approached and solved. By putting more than 50 tried-and-true techniques in writing, organized into mathematical approaches, physical/mechanical approaches, visual/graphic approaches, and abstract/conceptual approaches, and by illustrative examples of how each technique can be used, the young engineer-to-be or young engineer new to practice is made patently aware of *how* to solve problems.

To keep what could easily become tedium from closing off one's mind, if not making one close the book, the writing style is kept as comfortable and conversational as possible. But readability doesn't sacrifice technical accuracy or sophistication. Throughout the book, the author has interspersed what he hopes are enlightening and fun historical contexts and anecdotes. Likewise, each chapter contains figures that attempt to be visually striking as much as enlightening and tables that either summarize or support key concepts.

Most important, it is the author's sincere hope that readers will find this a joy-filled journey, as much as or more than simply another important technical book. Get ready to see the myriad of timeless and time-tested techniques used by our proud forbearers in engineering, and enjoy the journey.

Robert W. Messler, Jr.

Engineering Problem-Solving 101

Introduction

CHAPTER 1

Engineers as Problem Solvers

A *PROBLEM* IS defined as "a question or situation that presents uncertainty, perplexity, or difficulty."[1] For as long as human beings have existed, we have, as a species, been faced with problems that need to be overcome— sometimes to survive, sometimes to be safe or secure or more comfortable, sometimes to extend our mobility and/or our range, and sometimes simply (but profoundly!) to make our presence on this planet and in this universe apparent and more meaningful over time.

We are not alone as problem solvers. Other animals, from ants to apes, also face problems, over a wide range of complexity, that need to be dealt with. But our fellow species on Earth deal with problems in the first three categories only, and in order: (1) to survive, (2) to be secure, or (3) to extend mobility and/or range. Only we humans have ever taken on problems to make our presence known across space and time. This is not surprising when one considers the Hierarchy of Needs first proposed by the Brooklyn-born American psychologist Abraham Maslow (1908–1970) in his *Theory of Human Motivation* in 1943. As shown in Figure 1–1, the three lowest levels of Maslow's hierarchical pyramid—in order, "Physiological Needs" (sometimes given as "Biological Needs" or "Survival Needs"), "Safety Needs" (sometimes given as "Security Needs"), and "Belonging Needs" (sometimes given as "Love and Relationship Needs")—are collectively referred to as "D [deficiency] needs," while the two highest levels—in order, "Esteem Needs" (sometimes "Self-Esteem Needs") and "Self-Actualization Needs"—are collectively referred to as "B [being] needs."[2] The former of Maslow's two highest needs, involving the need to be accepted, respected, and valued by others, might be seen as shared by some higher-order mammals, as with *alpha males* among most herd or pack animals and among all primates. The latter, highest-of-all need, however, to be all that we are capable of being, is uniquely human.[3]

As but a few examples of animals solving problems are the following:

- Orangutans[4] and chimpanzees using sticks as tools to collect insects from holes and, as shown in Figure 1–2, by wild gorillas as stabilizers while they forage aquatic herbs—to eat and survive (Level 1).
- Orangutans using large leaves as umbrellas or rooftops to protect themselves (and their young) from the rain, as shown in Figure 1–3, or fern branches to protect themselves from the hot sun of Borneo—to be more secure and comfortable (Level 2).
- Beavers moving mud and rocks and cutting down small trees and branches to make dams and lodges (Figure 1–4)—to be more secure and comfortable, as well as to satisfy needs for relationships with mates and offspring by building communal lodges (Levels 2 and 3).

---

[1]*The American Heritage Dictionary of the English Language,* 4th edition, 2006, Houghton Mifflin Company, Boston, MA.
[2]Maslow subsequently extended his ideas to include his observations of humans' innate curiosity, which, especially apropos to the subject of this book, brings with it a need to be strong problem solvers.
[3]Viktor Emil Frankl, the Austrian neurologist and psychiatrist, added a sixth level to Maslow's hierarchy, i.e., "Self-Transcendence," in which people seek to go beyond their prior form or state.
[4]*Orang-utans* literally means "person" (*orang*) of the "forest" (*hutan*) in the native language of Borneo, where these great apes are found.

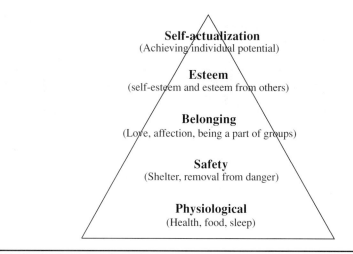

Figure 1-1   Maslow's Hierarchy of Needs.

- Beavers cutting and moving trees to create bridges across boggy areas and streams—to increase their mobility, extend their range, and attend to their needs for love, affection, and belonging by improving chances to find a suitable mate (Level 3).
- Ants building bridges using their own bodies to enable other ants to cross an obstacle (see YouTube.com, "Ants Engineering—Building Bridge Within No Time")—to increase their mobility and extend their range (Levels 1 and 3). Remarkable video, more remarkable problem-solving, as one watches the first ants to encounter the obstacle assessing the problem and a solution!

Of course, most of us are aware how clever common tree squirrels are at problem-solving when it comes to getting food from hanging bird feeders, even "squirrel-proof" ones, by causing them to spill seeds onto the ground by shaking or even spinning the feeder.[5] Recent studies have also proven that Aesop's fable "The Crow and the Pitcher," in which a crow drops pebbles into a pitcher half filled with water to raise the level to allow him to drink, has a real basis in fact. Rooks (relatives of crows) have been observed dropping small rocks into water to raise the level to allow a worm floating on the surface to be caught.[6]

Clever as our animal friends are, they don't come close to solving the number, diversity, and complexity of problems that we humans have solved and continue to solve since our appearance on this planet. In fact, there are people who are specially trained to solve problems for the benefit of others. They are known as *engineers*. According to the dictionary,[7] *engineers*, as a plural noun, are defined thus: "[Engineers] apply scientific and mathematical principles to practical ends such as the design, construction, and operation of efficient and economical structures, equipment, and systems." As a verb, *to engineer* means "to plan, manage, or put through by skillful acts or contrivance," the latter term meaning to "devise with cleverness or ingenuity." By this last definition, some of our animal friends are definitely Nature's engineers. But we human engineers truly excel at problem-solving.

No one could begin to list, no less describe, the innumerable problems solved by engineers over the ages, and this book is not intended to see what problems engineers have solved; rather, it is intended to compile and describe what techniques engineers use in problem-solving. But we can probably agree that problem-solving for engineers falls into one of two broad categories: *Problem-solving of necessity*

---

[5] Search YouTube.com under "Squirrels get to bird feeder . . ." to see real animal ingenuity at work!
[6] Check out www.sciencedaily.com/releases/2009/08/090806121754.htm.
[7] Ibid., Footnote 1.

Figure 1-2 A wild gorilla using a dead stick (top image, to her left) as a stabilizer while using her other, free hand to gather aquatic herbs. (*Source:* Originally by T. Breuer, M. Ndounoku-Hockemba, and V. Fishlock, in "First Observations of Tool Use in Wild Gorillas," *PLoS Biol* [*Journal of the Public Library of Science*] 3(11):e380, 2005; WikiMedia Commons from an image search, used freely under Creative Commons Attribution 2.5 license.)

and *necessary problem-solving.* The former involves solving problems that must be solved on their own account—as ends in themselves—the most obvious examples being problem situations that are life-threatening (e.g., preventing life-destroying flooding along rivers using dikes or protecting ships and sailors from life-threatening rock outcroppings using lighthouses). The latter involve solving the many and varied problems that arise in the course of voluntarily undertaking each and every design or construction project (e.g., stabilizing the soil for the foundation of a large civil structure or lifting heavy structural elements into place) or operating equipment, a process, or a system (e.g., achieving desired control for safety, efficiency, and economy). The Seven Wonders of the Ancient World are an interesting case in point.

Created in various versions in the first and second centuries BC, most notably by Antipater of Sidon and an observer identified as Philon of Byzantium, the original list was more a guidebook of "must see" sites for Hellenic tourists than any attempt to catalog great works of engineering located around the

Figure 1-3 Orangutans use large leaves as umbrellas or rooftops to improve their comfort. Here an orangutan in Tanjung Putting National Park near Camp Leakey in Kalimantan, Indonesia, uses leaves as protection from rain. (*Source:* Photograph taken by Paul and Paveena McKenzie and used with their kind permission.)

Figure 1-4 Beavers move mud and rocks and cut down branches and trees to build dams and lodges for security and comfort. (*Source:* www.furtrapper .com; O. Ned Eddins, Afton, WY, used with his kind permission.)

TABLE 1-1	The Seven Ancient Wonders of the World				
Wonder	**Construction Date**	**Builder**	**Notable Feature**	**Destruction Date/Cause**	**Modern Location**
Great Pyramid	2584–2561 BC	Egyptians	Tomb for Pharaoh Khufu	Still exists	Giza, Egypt
Hanging Gardens of Babylon	~600 BC	Babylonians	Multilevel gardens, with irrigation system. Built for Amytis of Media by husband Nebuchadnezzar II	~1st century BC/Earthquake	Al Hillah, Babil, Iraq
Temple of Artemis	c. 550 BC	Lydians, Persians, Greeks	Dedicated to Greek goddess Artemis; took 120 years to build AD 262	356 BC/by Herostratus Turkey by Goths	Near Selcuk, Izmir,
Statue of Zeus at Olympus (Statue)	466–456 BC (Temple) 435 BC	Greeks	Occupied whole width of temple aisle (12 m/40 ft)	5th–6th centuries AD/ Fire	Olympia, Greece
Mausoleum of Halicarnassus	351 BC	Carians, Persians, Greeks	45 m (150 ft) tall, adorned with sculptural reliefs	by AD 1494/ First a flood, then an earthquake	Bodrum, Turkey
Colossus of Rhodes	292–280 BC	Greeks	35 m (110 ft) statue of Greek god Helios	226 BC/ Earthquake	Rhodes, Greece
Lighthouse of Alexandria	c. 280 BC	Hellenistic Egypt, Greeks	115–135 m (380–440 ft) tall; Island of Pharos became the Latin word for *lighthouse*	AD 1303–1480/ Fires and earthquakes	Alexandria, Egypt

Mediterranean Sea.[8] Table 1–1 lists the Seven Ancient Wonders and some basic information on each, while Figure 1–5 shows an artist's concepts of the seven. A strong argument could be made that only the Lighthouse at Alexandria was probably a solution to a problem (for navigators to the greatest city of the time), that is, involved problem-solving of necessity. The others—while each a marvel of engineering at any time, no less during ancient times—were more an appeal to one of Maslow's two highest-level needs for self-esteem and self-actualization. The Hanging Gardens of Babylon, on the other hand, was surely created out of the need for King Nebuchadnezzar to express his love for his wife, Amytis of Media, fulfilling a Level 3 relationship need.

[8]For the ancient Greeks, the region surrounding the Mediterranean Sea was their known world (i.e., the Hellenic world). They probably didn't venture far beyond this area.

Figure 1-5 Depictions of the seven wonders of the ancient world by sixteenth-century Dutch artist Marten van Heemskerck: (*top, left to right*) Great Pyramid, Hanging Gardens of Babylon, Temple of Artemus, and Colossus of Rhodes; (*bottom, left to right*) Mausoleum of Halicarnassus, Statue of Zeus at Olympus, and Lighthouse at Alexandria. (*Source:* WikiMedia Commons from image search, originally contributed by mark22 on 7 May 2007; used freely under Creative Commons Attribution 2.5 license, but also within public domain.)

Two ancient accomplishments of engineering merit special attention here as examples of problem-solving of necessity versus necessary problem-solving: the Tunnel of Samos (also known as the Tunnel of Eupalinos) and the Great Pyramid of Giza, respectively.

Unquestionably, one of the greatest engineering achievements of ancient times is a water tunnel measuring 1036 m (4000 ft) in length, excavated through a mountain (Mount Kastro/Castro) on the Greek island of Samos in the sixth century BC. It was built of necessity to move fresh water from a natural source on one side of the island to a thriving city situated at the edge of a safe and busy harbor on the opposite side of the island—with the mountain separating the two sites. The marvel is that it was dug through solid limestone by two separate teams advancing in a straight line from opposite ends, using only hammers and chisels and picks (Figure 1–6).[9] The prodigious feat of manual labor is surpassed only by the mind-boggling intellectual feat of determining the direction of the tunnel so that the two digging parties met under the middle of the mountain. How this was done no one knows for sure, because no written records exist. But, at the time the tunnel was dug, the Greeks had no magnetic compass, no surveying instruments, no topographical maps, nor even much written mathematics at their disposal.

There have been some convincing explanations for how the tunnel was created, the oldest being one based on a theoretical method devised by Hero of Alexandria five centuries after the tunnel was completed. It called for a series of right-angled traverses around the mountain, beginning at one entrance of the proposed tunnel and ending at the other, maintaining a constant elevation. By measuring the net distance traveled in each of two perpendicular directions, the lengths of the two legs of a right triangle would be determined, with the hypotenuse of the triangle being the proposed path of the tunnel. By laying out smaller similar right triangles (see Chapter 8) at each entrance, markers could be used by each digging crew to determine the direction of tunneling. But how did the ancient Samosians keep their orthogonal straight-line segments at a constant elevation?

Tom M. Apostol, Emeritus Professor of Mathematics at CalTech, studied the problem for years and took issue with Hero's proposal. Professor Apostol's theory is detailed in a fascinating article, "The Tunnel of Samos," and was the subject of one of Dr. Apostol's wonderful lectures in Project MATHEMATICS on public-access TV (www.projectmathematics.com). The interested reader is strongly

[9]The English Channel Tunnel, like many other modern tunnels, was bored from opposite ends—one end by an English team, the other by a French team. Their tunnel segments met, but they had much more sophisticated technologies at their disposal.

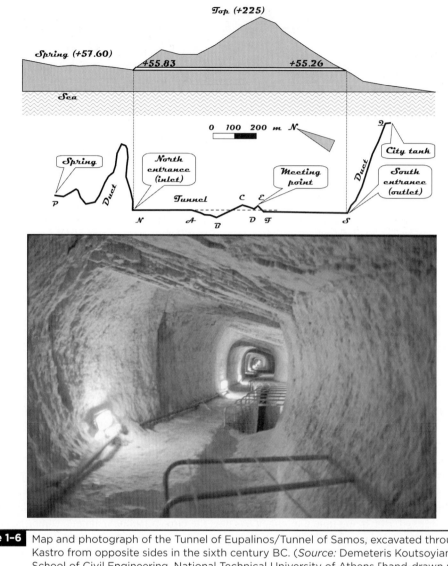

Top (+225)

Spring (+57.60)

+55.83   +55.26

Sea

0 100 200 m *N*

Spring   *North entrance (inlet)*   *Meeting point*   *City tank*   *South entrance (outlet)*

Duct   *Duct*

P   *Tunnel*   C E

N   *A*   B   *D F*   *S*

Figure 1-6 Map and photograph of the Tunnel of Eupalinos/Tunnel of Samos, excavated through Mount Kastro from opposite sides in the sixth century BC. (*Source:* Demeteris Koutsoyiannis, School of Civil Engineering, National Technical University of Athens [hand-drawn map at top] and WikiMedia Commons from image search [photograph at bottom], originally contributed by moumonza on 24 November 2007, used freely under Creative Commons Attribution 2.5 license.)

encouraged to read more about this amazing engineering accomplishment or, better yet, to watch the amazing series available on DVDs.

However it was done, the Tunnel of Samos stands as testimony to the extraordinary problem-solving skill of the ancient Samosians—born of necessity for water for survival.

While not born of necessity, but rather of faith and devotion (clearly higher-level needs!), the Great Pyramid, also known as Khufu's Pyramid, oldest and largest of the three pyramids in the Giza Necropolis bordering the Nile River near what is now Cairo, Egypt, in Africa, is probably the greatest testimony to the problem-solving ability of human beings ever (Figure 1–7)! Believed to have been constructed over a 20-year period concluding around 2560 BC, it was to serve as the eternal resting

Figure 1-7 Photograph of the Great Pyramids of Giza at the Giza Necropolis near Cairo, Egypt. (*Source:* Wikimedia Commons; originally contributed by Riclib on 17 June 2007, used freely under Creative Commons Attribution-Share Alike 2.0 Generic license.)

Figure 1-8 A CATIA rendering of the Great Pyramid of Khufu near the end of its construction over a 20-year period in which tens of thousands of workers moved a huge stone into place once every 2 minutes for 10 hours a day for 365 days a year. Also shown, surrounding the pyramid are living quarters for workers and staging areas for construction materials. (*Source:* Dassault Systemes from the 2011 edition of *Khufu Reborn*; used with the kind permission of Dassault Systemes.)

place for the mummified body of the beloved Fourth Dynasty Egyptian pharaoh popularly known as Cheops. The enormity of the Great Pyramid has posed—and continues to pose—a challenge to anyone who would attempt to explain how it could have been constructed—an engineering mystery that is delved into later in this book (see "Closing Thoughts").

It is estimated that this largest, essentially solid, pyramid consists of more than 2.4 million stone blocks averaging 2.27 metric tons (2.5 tonnes) each, along with additional blocks lying deep within the pyramid (where they served special functions) weighing much more. The overall height of the pyramid, when new, measured 146 m (480.9 ft). To accomplish building the Great Pyramid, one block would have to have been set in its final resting place every 2 minutes for 20 years . . . 365 days a year, working 10 hours a day! To add further to the imponderable questions of actual construction, there are the time and effort involved in cutting that many large stones from a quarry located 12.8 km (8 mi) away, and transporting them to the specially prepared 5.2-hectare (13-acre) construction site, a very early example of "just-in-time" manufacturing, in which stones were quarried and transported just before they were needed at the construction site. Tens of thousands of multiton white limestone facing stones were quarried more than 100 km (60 mi) away, and also had to be transported. Theories abound, but no irrefutable solution has been found. (See Figure 1–8.)

Table 1–2 lists some dimensions for the Great Pyramid that, with a little reflection, are also staggering.[10]

The dimensional accuracy of the lengths of the four sides of the square base is ±0.20 m (±0.67 ft), or ±0.044 percent—and this is the accuracy over sides that are 230.5 m (756 ft) long! The angular accuracy

[10]The first to precisely measure the Great Pyramid was W. M. Flinders Petrie, who published his findings in a book (*The Pyramids and Temples of Gizeh*) by the Royal Society of London in 1883. The Great Pyramid was professionally surveyed by J. H. Cole in 1925, with his measurements being published in *Determination of the Exact Size and Orientation of the Great Pyramid*, by Government Press, Cairo.

TABLE 1-2 Key Dimensions of the Great Pyramid*	
Height (including capstone):	146.55 m (480.69 ft)
Courses (of stones):	201
Length of sides (at base): West: North: East: South:	230.42 m (755.76 ft) 230.32 m (755.41 ft) 230.45 m (755.87 ft) 230.51 m (756.08 ft)
Perimeter:	921.70 m (3023.22 ft)
Angle of corners: Northwest: Northeast: Southeast: Southwest:	89° 59´ 58˝ 90° 3´ 02˝ 89° 56´ 02˝ 90° 3´ 02˝
Slope (of north face)	51° 50´ 40˝
Area of base:	4,050 m² (4,846 yd²) 0.405 hectares (13 acres)

* Data were selectively taken from measurements recorded by Peter Lemesurier in his 1977 book, *The Great Pyramid Decoded,* reprinted by Elements Books Ltd., 1996.

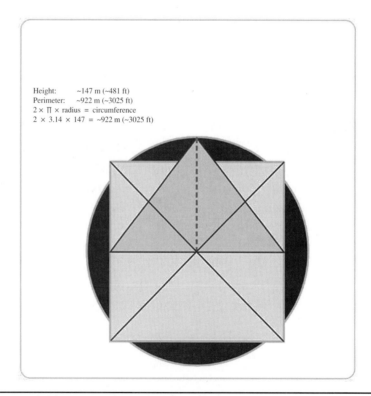

 Figure 1-9 The remarkable Pi ratio of the Great Pyramid, in which the base of the Great Pyramid is a square whose perimeter is equal to the circumference of a circle with a radius equal to the height of the Great Pyramid.

Figure 1-10 The 102-story Empire State Building in New York City. Completed in 1931, at 381 m (1250 ft), it was the tallest building in the world until the World Trade Center "Twin Towers" in New York City was completed in 1972. (*Source:* WikiMedia Commons from an image search, originally contributed by Jinguang Wang on 31 May 2010, used freely under Creative Commons Attribution 2.5 license.)

of the four corners of the square base is +0.048°, or 0.054 percent. This linear and angular degree of accuracy would absolutely challenge modern construction companies.

It turns out, albeit not coincidentally, that the sides of the base of the Great Pyramid create a square whose perimeter is almost exactly the same as the circumference of a circle whose radius is the height (or altitude) of the pyramid (Figure 1–9). Thus, 2π times the radius (here, the pyramid's altitude), which is the circumference of a circle, compares to the perimeter of the pyramid's square base as: 2(3.14) (480.69 ft) = 3020.68 ft versus 3023.22 ft. All this was accomplished 2200 years before Euclid's treatise on geometry was completed in 350 BC.

The real marvel may be that the Great Pyramid of Giza was undertaken out of love and affection for a pharaoh who sought to fulfill a need for his own self-esteem. In the process, many, if not all, of those who designed and built the Great Pyramid surely achieved self-actualization.

Before moving on to the purpose of this book—that is, to present the time-tested and timeless techniques engineers use to solve problems—it is worth considering some wonders of the modern world in which we live, as these serve well to bring to mind the innumerable problems that were necessary to solve in the course of creating each wonder.

The American Society of Civil Engineers (ASCE), with the help of experts from around the world, selected Seven Wonders of the Modern World. These are:

- The Empire State Building in New York City (Figure 1–10)
- The Itaipu Dam spanning the border between Brazil and Paraguay, South America (Figure 1–11)
- The CN Tower in Toronto, Ontario, Canada (Figure 1–12)
- The Panama Canal in Panama, Central America (Figure 1–13)

Figure 1-11 The Itaipu Dam on the border between Brazil and Paraguay is the largest operating hydroelectric facility in terms of power output at 94.7 TWh in 2008 and 91.6 TWh in 2009 (versus Three Gorges at 80.8 TWh in 2008 and 79.4 TWh in 2009). (*Source:* WikiMedia Commons from an image search, originally contributed by Cyc on January 2005, used freely under Creative Commons ShareAlike 1.0 license.)

Figure 1-12 The CN (Canadian National) Tower in downtown Toronto, Ontario, Canada; the communication and observation tower was the tallest and is now the fourth-tallest freestanding structure in the world at 553.3 m (1815 ft). (*Source:* WikiMedia Commons from an image search, originally contributed by wladyslaw on 10 September 2008, used freely under Creative Commons Attribution 2.5 license.)

- The Channel Tunnel between Folkestone/Kent, England, and Calais, France (Figure 1–14)
- The North Sea Protection Works in The Netherlands (Figure 1–15)
- The Golden Gate Bridge in San Francisco, CA (Figure 1–16)

It is left to the reader to decide what category of problem was being solved. Regardless, each is impressive!

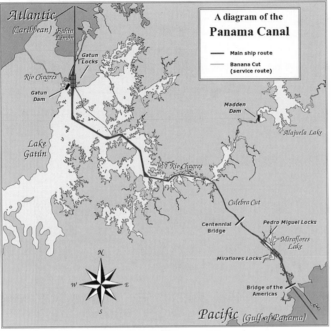

(a)

(b)

Figure 1-13 The Panama Canal, built by the U.S. Army Corps of Engineers between 1904 and 1914, links the Atlantic Ocean and Pacific Ocean through the Isthmus of Panama (a) with a series of canals and lakes totaling 77 km (48 mi), dramatically shortening the time required for ships to travel from one ocean to the other (b). (*Sources:* WikiMedia Commons from an image search, [a] originally contributed by johanheghost on 25 November 2005; [b] originally contributed by Dozenost on 27 May 2005, used freely under Creative Commons Attribution 2.5 license.)

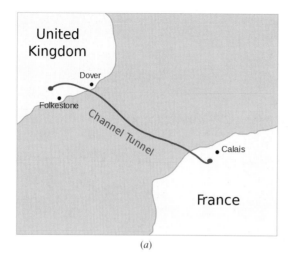

(a)

(c)

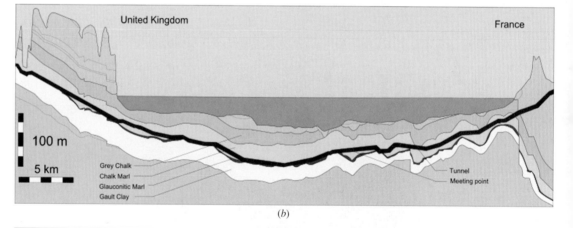

(b)

Figure 1-14   The [English] Channel Tunnel—or "Chunnel"—links England to continental Europe with a 50.5 km (31.4 mi) train tunnel. It includes the longest undersea tunnel in the world, at 37.9 km (23.5 mi). A map of the tunnel's location (a), a simple geological profile (b), and an exhibit of a section of the tunnel and a EuroStar train at the National Railroad Museum in York (c) are shown. (*Source:* All from Wikimedia Commons; map originally contributed by Montadelo2005 on 21 May 2007 under Creative Commons Attribution 3.0., profile originally contributed by Commander Keane on 21 January 2008 under Creative Commons Attribution Share-Alike 3.0 Unported, and section of tunnel originally contributed by Commander Keane on 8 January 2008 under Creative Commons Attribution 2.5 Generic, all used under free license.)

(a)   (b)

Figure 1-15 The North Sea Protection Works is a marvel of engineering to control the power of the sea, creating arable land for the Dutch people in The Netherlands; see satellite view (a). The system consists of a vast and complex assortment of dams, floodgates, swing-away storm surge barriers, as at the Maeslant Barrier (b), and other structures. (*Source:* [a] from WikiMedia Commons from NASA's public-domain global software, 21 October 2005, for which no permission for use is required; [b] from www.DigiPhot .nl/luchlfoto/1200px_c/050404-118-sturmvloedkering.jpg, used with permission from DigiPhot.)

Finally, for further reflection, in 2003, the National Academy of Engineering in the United States published *A Century of Innovation: Twenty Engineering Achievements that Transformed Our Lives.* The list follows:

1. Electrification
2. Automobile
3. Airplane

Figure 1-16 The Golden Gate Bridge in San Francisco spans the entrance to the San Francisco Bay off the Pacific Ocean. Its 1280-m (4200-ft) main span was once one of the longest in the world but is now dwarfed by newer suspension bridges. It is, however, one of the most recognizable and aesthetically pleasing suspension bridges in the world. (*Source:* WikiMedia Commons from an image search, originally contributed by Rich Niewiroski, Jr. [www.projectrich.com] on 1 January 2007, used freely under Creative Commons Attribution 2.5 license.)

4. Water supply and distribution
5. Electronics
6. Radio and television
7. Agricultural modernization
8. Computers
9. Telephone
10. Air-conditioning and refrigeration
11. Highways
12. Spacecraft
13. Internet
14. Imaging
15. Household appliances
16. Health technologies
17. Petroleum and petrochemical technologies
18. Laser and fiber optics
19. Nuclear technologies
20. High-performance materials

Again, it is left to the reader to decide what the motivation for each may have been. But who can disagree that each involved extraordinary problem-solving skills?

Now let's get on with seeing what techniques engineers use to solve problems—the real purpose for this book.

Suggested Resources

Video/DVD

Aspotol, Tom M., *Tunnel of Samos* (videotape or DVD), Project MATHEMATICS! Series, 1988–2000.

Books

A Century of Innovation: Twenty Engineering Achievements That Transformed Our Lives, National Academies Press, Washington, D.C., 2003.

Clayton, Peter A., *The Seven Wonders of the Ancient World,* edited by Peter A. Clayton and Martin Price, Routledge/Taylor & Francis Group, London, 1988.

Jackson, Kevin, and Jonathan Stamp, *Building the Great Pyramid,* Firefly Books Ltd., Toronto, Ontario, Canada, 2003.

Maslow, Abraham, *Motivational Personality*, 3rd edition, Harper & Row, New York, 1954.

Parklyn, Neil, *The Seventy Wonders of the Modern World,* Thomas & Hudson Publishing, Los Angeles, 2002.

CHAPTER 2

Problem-Solving Skills versus Process versus Techniques

HOPEFULLY, now that you know that engineers earn a living by solving problems, you appreciate that problem-solving is an essential skill for engineers at all stages of their learning—as undergraduate students in school and as engineers in practice.[1] In fact, having strong problem-solving skills is often the key differentiator in career success for engineers. The best problem solvers tend to advance the fastest and the furthest. The problems to be solved may arise internally (e.g., during the process of design, from in-house manufacturing or construction processes and/or processing, or during tests) or externally (e.g., in response to the request of a client for a new or improved product, from unforeseen complications from material or part suppliers, or as part of supporting a product throughout its service life). Such problems can be small or large, simple or complex, and easy or difficult to solve. But, regardless of their nature, an engineer's job is to solve them. Thus, being a capable and confident problem solver is an extremely important skill to be developed early, versus late, in an engineer's training.

Confidence for solving problems comes from three things: (1) having a good process for tackling problems whenever, however, and wherever they arise; (2) having an array of techniques to employ under different circumstances and in different situations; and (3) having experience (Figure 2–1). As undergraduate engineering students, you don't have much control over gaining experience because the best experience comes only with time. (Hence, the adage: "Time is a great teacher . . . but, unfortunately, it kills all its pupils," Hector Louis Berlioz [1808–1869], French Romantic composer.) Similarly, there is only a limited amount that teachers in engineering can do to impart experience because their time with each student is limited and, increasingly so, they themselves have little or no practical experience with problem-solving as working engineers. It should come as no surprise that the best way to learn is to do.

A historical approach has been to immerse engineering students in problem-solving by assigning homework problems and giving tests.[2] The hope, especially with homework as preparation for tests, is that by having to solve problems, students will develop problem-solving skills. This may or may not be true. Surely it varies greatly from student to student. The analogy to throwing someone into deep water to have them learn to swim—after some lectures on the theory of swimming, of course—is risky, to say the least. Some will drown. Some will swallow water, sink, and, once rescued, may well decide that swimming is not for them. Hence, there are more dropouts from undergraduate engineering than the United States or the world can afford, or, worse, than might be necessary with a different approach.

The principal reason this approach has worked at all is that the great majority of problems assigned by engineering academicians involve what are called "plug-and-chug" problems. Such problems involve

---

[1]Engineers are not the only ones who need to solve problems, as doctors do it, lawyers do it, businesspeople do it, "even the birds and bees do it," to paraphrase Cole Porter's lyrics in his "Let's Do It (Let's Fall in Love)," and as can be seen by examples in Chapter 1.

[2]Think of how many problems an undergraduate in engineering has to solve to survive one or more courses in chemistry, physics, and mathematics, as well as courses in heat and fluid transfer, thermodynamics, statics, dynamics, etc., and within the student's major.

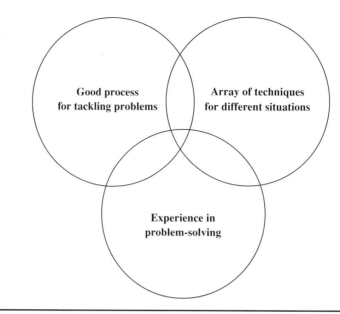

Figure 2-1 Schematic showing the keys to strong problem-solving skills.

using a specific mathematical equation (usually stipulated in the assignment), plugging in given values for each known variable, and chugging away at the math until the unknown variable is calculated. The reality is, there are many problems encountered by engineers—arguably, most problems—that are not nearly so straightforward as to require solution of a single equation when provided with all of the needed inputs.

For problems more complex than simple plug-and-chug types, even when a student succeeds by arriving at a solution, it only shows that he or she can solve *that* problem. If the student isn't aware—or is not explicitly made aware—of what technique(s) he or she used to arrive at the solution, it is questionable whether he or she would do well on even a similar problem. Thus, the idea that posing large numbers of problems, even of a wide variety of types (beyond plug-and-chug), requiring a wide variety of problem-solving techniques (beyond precise mathematics), in a wide array of subject areas (i.e., courses), improves problem-solving skill is questionable. It is inefficient, at least, as time might be better spent, and may well be flawed, at worst, if the approach either fails or leads to less capable—or fewer—problem solvers.

Problem-Solving Skills

Having worked as an engineer for a living before teaching engineering, the author sees three serious shortcomings in educating young engineers that need to be remedied to best develop problem-solving skills. First, the kind of problems students work to solve—in all courses, from the basic sciences and mathematics to preengineering through discipline-specific courses—should be selected and/or structured to explicitly teach and develop specific problem-solving techniques, beginning with the most basic techniques and moving on to more complex techniques as well as combinations of techniques. Also, problems posed should not all be suitable to mathematical solution alone. Problems should be posed that require physical or mechanical approaches (e.g., such as mechanical dissection), graphic approaches, and abstract approaches (e.g., decision trees). Second, problems posed should evolve from straightforward closed-ended (or closed-solution) types, for which there is only a single, specific solution, to open-ended types, for which there are many different possible solutions, depending on specific needs. Many of these kinds of problems require nonmathematical approaches for their solution. Third, as the student advances, problems should be posed the way they usually are posed in practice, that is, somewhat nebulously, to force the student/engineer to formulate the essential problem himself or herself and without providing exactly what is needed for the solution. This includes not making it always obvious what equation to use, for example, and/or not providing all or just the needed inputs.

Working backward, from the third to the first stated shortcoming, here is the author's suggestion for developing problem-solving skill:

- To the third shortcoming, learn, as students (teach, as instructors or mentors), how to find the information/data needed to solve a problem if that information/data is/are not provided in the initial problem statement or, alternatively, learn to select just what is needed from among more information/data than is/are needed if that is the case. After all, what does one really learn from plugging in given values for mass m and acceleration a in Newton's equation $F = ma$ to find the force F? Realistically, problems encountered by engineers are more complicated than this. Sometimes values for needed inputs are not immediately at hand, even when a relevant equation is known. Other times there is no obvious equation, even though there is a tremendous amount of data, albeit of questionable value.

  As an example of the former situation, think of what it would take to calculate the maximum distance a baseball could be hit if all that one knew were the mass of the baseball, the mass of the bat, the velocity of the pitched baseball, and the velocity of the swung bat, even assuming the ideal initial launch angle of the hit baseball. The problem is much more complex than to assume conservation of momentum and conversion of initial velocity into distance against the force of gravity. How much energy is lost due to deformation of the elastic (versus perfectly inelastic) baseball? How much energy is lost due to elastic deformation (due to deflection) of the bat? What is the effect of air resistance? And so on. As an example of the latter situation, think of the difficulty of picking the horse that will win the Kentucky Derby based on the extensive information/data available on each horse—sire/father and dam/mother (i.e., breeding), trainer, jockey, purse money won, number of first-, second-, and third-place finishes, preference for dirt or grass, preference for short or long races, practice times, and so on. What equation—or "handicapping system"—should be applied to predict the outcome? This, of course, assumes that only such information/data influences performance, and ignores how the horse feels. There is likely no predictive equation, only probability.

- To the second shortcoming, closed-ended (or closed-solution) problems, for which there is one and only one solution, should be used to learn (teach) techniques and hone skill by moving from simple to more complex problems, while open-ended problems should be used to force (teach) students how to weight various options among possible alternatives. As an example of a closed-ended engineering problem: How much thrust is required to place a satellite of known mass into a specific orbit around the Earth? This problem is actually more complicated than it might first appear when one recognizes that more thrust to deal with a heavier satellite or to go into a higher orbit requires a larger rocket, which requires—in and of itself—more thrust to move the rocket. An example of an open-ended problem is: What approach would allow like-colored as well as different-colored balls of two different sizes (e.g., about the size of a golf ball and about the size of a tennis ball), some of which are solid and some of which are hollow, to be collected, sorted, and returned to storage racks in a playroom after they have been scattered around the room by toddlers? Engineers would be inclined to explore various approaches from (1) inflatable floor mats to force balls to roll into troughs to be lifted by conveyors to then roll down ramps with sized holes (for the small versus large balls to drop through) and, then, sort by weight (i.e., solid versus hollow) using the ball's trajectory off the end of a ramp once sorted by size to (2) sophisticated robots with vision systems. In the meantime, one team of engineering students at the author's university assigned this problem ended up having one student's mother brought out at the final presentation to show how she could collect, sort, and return the balls to storage racks. This may not have been what was expected by the engineering professors from the engineering students, but it was certainly a viable solution.

The key with open-ended problems is to teach students of engineering how to weigh options for the solution that is good enough, not necessarily best—but *only* after the problem has been unambiguously stated or formulated!

■ To the third shortcoming, this book focuses on the techniques that engineers have used over time to successfully solve problems.

The Problem-Solving Process

There are dozens of books written on the process of (or strategies for) solving problems within engineering, predominantly during design, which often involves creative problem-solving. Some are quite good, and all ultimately divide the process into four steps. Without going into detail here, these four steps (shown in Figure 2–2) are:

Step 1:   Defining (or formulating) the problem
Step 2:   Generating alternatives (as possible solutions)
Step 3:   Evaluating and selecting among any alternatives
Step 4:   Implementing the chosen solution

The reader interested in the *process* by which problems are solved is encouraged to seek other references, some of which are listed at the end of this chapter. But there are a few points that need to be made here.

It is important to have a systematic process for several reasons, not the least of which is to allow oneself or others to retrace steps if a fully satisfactory solution is not found. Having a process also provides a framework to guide thinking, which you will later see is very important.

No step is more important than Step 1—defining the problem, by which is meant *properly* and *fully* (i.e., unambiguously) defining the problem. If the problem is not properly identified and fully defined (or formulated), the desired solution may not be found. Going back to the example of the problem of collecting, sorting, and returning balls to storage racks, cited earlier, by not stipulating "a device," the

Step 1:   Defining (or Formulating) the Problem
  o Identifying the real problem to allow a root-cause
     solution
  o Fully articulating what constitutes success

Step 2:   Generating Alternatives as Possible Solutions
  o Keeping an open mind to all possibilities
  o Eliciting creativity as appropriate

Step 3:   Evaluating and Selecting Among Alternatives
  o Remaining objective
  o Weighting alternatives for both tangible and
     intangible factors
  o Applying necessary and/or appropriate constraints

Step 4:   Implementing the Chosen Solution
  o Checking the validity and viability of the chosen
     solution
  o Following up on the success of the solution

Figure 2-2 Summary of the four steps comprising an effective process for solving problems, most notably, but not solely, in design.

opportunity for the unintended solution of having someone's mother do it can result as a suitable option as "an approach."

Spending time to clarify and define the problem also serves two other important functions: First, it helps one get his/her brain in gear—in essence, to help begin the process of finding a solution. Our brains need help. We need to focus our thinking. Taking time to think through the problem being faced to properly and fully define it—as opposed to "jumping right in [to the pool]"—helps one's brain begin to collect, generate, and organize thoughts. Second, a properly defined problem also allows one to find a solution (if not *the* solution) to the *root-cause* of the problem. Engineers need to find root-cause-based solutions to preclude recurrence of the problem. An example may help.

If some people walking down an aisle in a workplace have slipped on a banana peel, it is critical to find the root-cause of this problem, which may appear to be the presence of the banana peel. Some engineers might be inclined to build a small bridge over the banana peel. This would certainly prevent anyone else from slipping on *that* banana peel, but it might not resolve the real problem, which is the appearance of banana peels where they do not belong. Without finding the root-cause for misplaced banana peels, one could easily end up with lots and lots of small bridges! The problem is not *the* banana peel. It is the unwanted presence of banana peels in aisle ways.[3]

While this book will not help you define a problem, it will (in Part 4) address the generation of alternatives for solving a problem. How to properly formulate a problem is treated well in almost any book on design.

The other steps of problem-solving should also be read about in other books.

Problem-Solving Techniques

Since no book has been written that specifically addresses the techniques (or approaches) engineers use to solve problems, one could debate what techniques (or approaches) such a book should include, no less how these techniques might be logically organized. For want of any better system of organization, the author, in collecting his thoughts for this book, chose to divide his list of techniques into four categories:

1. Mathematical approaches
2. Physical/mechanical approaches
3. Visual/visual-aid approaches
4. Conceptual (or thought-based) approaches

Mathematical approaches all share a basis in (or, at least, a clear tie to) mathematics or mathematical principles. Many of these approaches, not surprisingly, involve manipulation of numbers, including plug-and-chug. But not all do. Some involve geometry without numbers. Engineering faculty who have not had practical experience in engineering in industry, for example, tend to focus almost exclusively on mathematical approaches when they teach students how to solve problems. Furthermore, in their bias toward (as the result of their comfort zone within) mathematics from their own formal engineering education, they also tend to choose problems that are suited to mathematical approaches in order to find a solution. They often ignore problems that require other nonmathematical approaches out of either lack of awareness of or familiarity with (some might say ignorance of) such techniques.[4] However, even for the host of problems most engineering faculty work out as examples, or assign as homework, or pose in tests, they do not identify the specific technique or approach being used. This is, unfortunately, left to the student to deduce.

---

[3]The piecemeal solution of an ill-defined problem is far from uncommon in engineering. In engineering, small "fixes" eventually lead to a need for an entirely new design. A wonderful book is that by Henry Petroski, *To Engineer Is Human: The Role of Failure in Successful Design,* Vintage Books/Random House, New York, 1992 (originally published in a slightly different form by St. Martin's Press in 1985).

[4]Some rigid academicians view nonmathematical approaches to problem-solving as illegitimate. They are wrong!

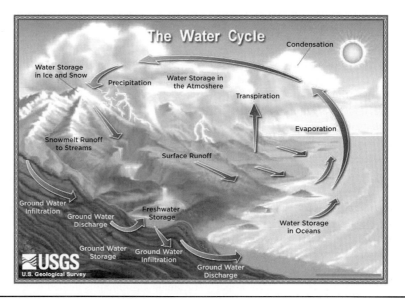

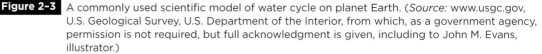

Figure 2-3   A commonly used scientific model of water cycle on planet Earth. (*Source:* www.usgc.gov, U.S. Geological Survey, U.S. Department of the Interior, from which, as a government agency, permission is not required, but full acknowledgment is given, including to John M. Evans, illustrator.)

Physical/mechanical approaches all share a basis in some physical object or construct and/or some mechanical process. These techniques tend to work quite well for those well suited to engineering (based on innate aptitude) because such people tend to like to touch things—taking them apart (like an old clock or Mom's vacuum cleaner) and/or putting them together (like Lego blocks). For students of engineering to learn—and for instructors or mentors to teach—such techniques is important because, after all, engineers deal and manipulate our (and their) physical world. Interestingly, our kindred spirits in architecture, even with the advent and availability of rich solid-modeling software, still employ physical models made from wood Popsicle sticks, wire coat hangers, and papier-mâché. So do engineers in industry.[5]

Visual/visual-aid approaches have in common the use of a visual medium and/or visual assessment. These techniques can include graphing, which might also be categorized as a mathematical approach. Such techniques tend to appeal most to those who could be classified as visual learners, but they clearly have value for the assessment of hard-to-quantify, less tangible/more aesthetic aspects of problem solutions (e.g., many designs) such as form, line, style, texture, visual appeal, and so on.[6] For such people, ideas, concepts, data, and other information are associated with images. By representing information spatially and with images, visual learners are able to focus meaning, reorganize and group similar ideas easily, and make better use of their visual memory. The aforementioned physical/mechanical approaches tend to appeal to both tactile and visual learners.

One will note that mock-ups and prototypes, covered in Part 3 on visual or graphic approaches, could

---

[5]The folding wing used on carrier-based Navy aircraft to allow their compact storage was developed for World War II by an engineer at the Grumman Aircraft Company from an initial physical model involving a pink rubber eraser (used by draftsmen) as the aircraft's fuselage and two bent paper clips as the wings. The paper clips could be rotated to show how the wing would "fold" up against the fuselage.

[6]The styles by which people learn (i.e., learning styles), according to a model proposed by Neil D. Fleming with Colleen Mills in 1992, include visual, aural (or auditory), and kinesthetic (or tactile) learning (i.e., the VAK model), later modified to also include read/write learning (i.e., in the VARK model).

Category of Problem-Solving Technique	Scientific Modeling Method
• Mathematical approaches	• Mathematical models
• Physical/mechanical approaches	• [No equivalent]*
• Visual/visual-aid approaches	• Graphical or iconic models
• Conceptual approaches	• Abstract or conceptual models

* In science, experimentation or experiments are used as a physical approach to understanding.

Figure 2-4 Schematic mapping of the four categories of problem-solving techniques used by engineers against the various methods used in scientific modeling.

have been (or could be) listed here in Part 2 on physical/mechanical approaches instead or as well. This simply points out possible different uses of the same technique during problem-solving.

Conceptual (or *thought*) *approaches* have in common that they all involve or partially draw upon or appeal to one's mind, although some are clearly visual (e.g., morphological charts and storyboarding). Many of the techniques listed under this category involve more abstract thinking (e.g., analogs and metaphors), but not all (e.g., storyboarding). Others involve ways to simplify (e.g., black boxes) or subdivide (e.g., decoupling) a complex problem or, later in the process of fully developing the solution, to combine or integrate ideas (e.g., coupling).

Some might argue for conceptual approaches to appear first in a book of problem-solving techniques. They do not because, even though most, if not all, problems do (or should) initially be approached by thinking as opposed to acting,[7] the techniques listed under this category are among the most intellectually sophisticated. Hence, they are addressed last, after less intellectually demanding techniques have been presented. This said: *All* problems should be approached first by thinking!

If one is familiar with—or becomes familiar with—scientific modeling, one will note, as the author was led to note, the parallel between the various methods used there to the categories of problem-solving techniques used here. *Scientific modeling* is the process of generating abstract or conceptual, graphical, and/or mathematical models, i.e., three of the four categories. Scientific models seek to represent empirical[8] objects, phenomena, and physical processes in a logical and objective way. All models are simplified reflections of reality.

Figure 2–3 presents a common scientific model for the water cycle. Figure 2–4 maps the categories of problem-solving techniques against scientific modeling methods.

Using This Book

As for how things are presented in the remainder of this book, each chapter (and its general technique or highly interrelated techniques) begins with a description of the technique(s) and ends with an illustrative example or two showing possible use of the technique(s).

Table 2–1 lists the various problem-solving techniques to be presented in this book. All have been tested by time and all those covered are timeless.[9]

---

[7] Remember, it may not be wise to jump into deep water before one can swim or know what is in the water!

[8] *Empirical* is defined as originating from or based upon observations or experience.

[9] For this reason (i.e., timelessness), the modern technique of computer-based solid modeling, as well as various computer-based simulation techniques and methods are not covered in any detail. The interested reader is encouraged to seek other references on these techniques. Quite simply, it is first and foremost a visual technique, albeit based on the mathematics of boolean algebra.

TABLE 2-1	A List of the Various Problem-Solving Techniques Available to Engineers

Mathematical Approaches
- Equations ("plug-and-chug")
- Approximating
- Estimating
- Interpolating and extrapolating
- Checking
- Dimensional analysis/reconciling units
- Using ratios
- Similarity
- Using indexes
- Scaling (in variant design)
- Sensitivity analysis (of parameters)
- Response curves and surfaces
- Numerical methods
- Dimensionless parameters
- Assumptions, conditions, and cases

Physical/Mechanical Approaches
- Reverse engineering (mechanical dissection)
- Material property correlations
- Making simple models (for proof of concept)
- Experimental models/experimentation
- Models for testing
- Mock-ups
- Prototypes and rapid prototyping
- Trial and error

Visual/Visual-Aid (Iconic) Approaches
- Sketching
- Tracing and transferring
- Loftings and lines taking
- Graphing and graphical methods
- Venn diagrams
- Layouts
- Fishbone diagrams or Ishikawa diagrams
- Flowcharts
- Templates
- Models, mock-ups, prototypes, and test vehicles

Conceptual Approaches
- Brainstorming methodology and techniques
- Analogs and metaphors
  - —Thermal-electrical analogs and vice versa
  - —Electrical-mechanical analogs and vice versa
  - —Biological analogs (bionics or biomimetics)
- Problem dissection/decoupling and coupling
- Backward or reverse solution
- Morphological charts
- Storyboarding/planning
- Using black boxes

As for where the engineering student or lifelong student of engineering should begin when posed with a problem, the simple answer is: Know what techniques you have at your disposal and how to use each one properly. The analogy to a mechanic knowing his or her tools is a good one. Once you know the various techniques available, which one to use in any specific situation will become more and more apparent with time and greater experience.

This book is a proper beginning, not the end, for an engineer. It must be read once by every student of engineering and will, throughout one's career as an engineer, undoubtedly be consulted over and over again.

Suggested Resources

Eide, Arvid, Roland Jenison, Larry Northrup, and Lane Marshaw, *Introduction to Engineering Design and Problem Solving,* 2nd edition, McGraw-Hill, New York, 2001.

Karsnitz, John R., John P. Hutchinson, and Stephen O'Brien, *Engineering Design: An Introduction,* Cengage Learning, Independence, KY, 2008.

LeBlanc, H. Scott, *Strategies for Creative Problem Solving,* 2nd edition, Prentice Hall, Upper Saddle River, NJ, 2007.

Petroski, Henry, *Invention by Design: How Engineers Get from Thought to Thing,* Harvard University Press, Cambridge, MA, 1996.

PART ONE

Mathematical Approaches to Problem-Solving

CHAPTER 3

Using Equations (Plug-and-Chug)

THE MOST OBVIOUS and most common, although not necessarily the simplest, technique for or approach to solving an engineering (or science) problem is to employ a relevant or pertinent equation, substitute numerical values for each known parameter or variable in the equation, and solve for the unknown parameter or variable. This technique simply requires appropriate arithmetic or mathematical manipulation to arrive at a single, closed-ended analytical solution.[1] The technique is commonly referred to by students and teachers alike as "plug-and-chug," for obvious reasons. In order for this technique to be applicable, the posed problem must involve the following characteristics:

- There must be a single quantitative solution to the problem.
- There must be an applicable or relevant mathematical equation.
- Quantitative (numerical) values are needed for each known parameter or variable in the pertinent equation.

If more than one parameter or variable are unknown, an equal number of equations that interrelate the common parameters or variables are required, and these must be solved simultaneously (i.e., simultaneous equations are needed). Two equations for two unknowns, three equations for three unknowns, and so on.

Effective application of mathematical equations to solve a problem requires consideration of three essential factors:

1. Units for all known and unknown (i.e., needed or desired) parameters or variables
2. Conversion factors to reconcile units within the equation and to the units required or desired in the solution
3. Significant figures in the solution

Besides these essential considerations, one may also require a value (or values) for any physical or mathematical constant in the equation. Finally, while not always necessary, it is often valuable and always instructive to consider the form of the equation.

Before moving on to an illustrative problem for problem-solving using plug-and-chug mathematical equations, let's consider the just-mentioned factors.

Units

Problems for which there is a quantitative solution involve units for each and every parameter in each and every applicable equation. All physical (real) quantities have units. *Units* fix the magnitude of the

---

[1] Such problems are generally referred to as *closed-ended problems* or *closed-solution problems*.

TABLE 3-1	Examples of English versus Metric Units for Some of the Most Important Dimensions in Engineering and Science	
Dimension	**Metric/SI Unit**	**English Engineering Unit**
Time	second (s)	second (sec)
Length	meter (m) centimeter (cm)* kilometer (km)	foot (ft) inch (in) mile (mi)
Mass	kilogram (kg) gram (g)*	pound mass (lbm)
Temperature	Celsius (°C)	Fahrenheit (°F)
Absolute temperature	kelvin (K)	Rankine (°R)
Force	newton (N) dyne (dyne)*	pound force (lbf)
Energy	joule (J) erg (erg)*	British thermal unit (Btu) calorie (cal)

*These units are the basis for the cgs (versus kgs) system.

physical quantity on some scale for some dimension. Noninclusive examples of important dimensions are length, time, mass, temperature, absolute temperature, force, a nd energy. Two common systems of units are used in science and engineering, i.e., English units and metric units.[2] While still in use (predominantly, but not exclusively, in the United States), English units have largely been—and will likely continue to be—supplanted by metric units.[3] The universal use of the metric system would greatly facilitate a global marketplace, in which there would, of necessity, be complete compatibility and interchangeability around the world.

Table 3–1 gives a few examples of English units versus metric units for the aforementioned dimensions. A more comprehensive list of units can be found on the Internet (searching for "units in engineering and science").

Before attempting to solve a mathematical equation for some unknown needed or desired parameter or variable, the units for all known parameters or variables must be inserted into the pertinent equation, rearranged as necessary to solve for the needed or desired unknown, *and* the units for the needed or desired solution must be known.

Illustrative Example 3-1: Units in an Equation

If one wished to calculate for how much time *t* some object would have to be accelerated at a constant rate *a* for the object to reach a given final velocity *v*, the initial operative mathematical equation is:

$$v = v_\text{o} + at$$

[2]Collectively known as *SI (Standard International) units*, metric units actually employ two subsystems known as mks and cgs. The former, generally used by engineers, has as its basis the meter (m), kilogram (k), and second (s), while the latter, used more often by scientists, has as its basis the centimeter (c), gram (g), and second (s). These systems become more complicated when energy, force, pressure, etc., are considered.

[3]One great preference for the metric system is that its units have a basis of 10. This is not the case for English units, which can have any of several bases (12 inches per foot, 16 ounces per pound, 14 days per fortnight, 8 furlongs per mile, etc.).

in which the object had an initial velocity v_o. If the initial velocity was zero (i.e., $v_o = 0$), the equation simplifies to:

$$v = at$$

Rearranging this equation to allow for time t to be found, gives:

$$t = v/a$$

If the time is desired in seconds (s), the SI units for v and a need to be m/s and m/s^2, respectively.

For given values of $v = 200$ m/s and $a = 9.8$ m/s^2 (the approximate value of the acceleration due to gravity on Earth), the time t would be:

$$t = (200 \text{ m/s})/(9.8 \text{ m/s}^2) = 20.4 \text{ s}$$

Conversion Factors

In order to generate a solution having the correct needed or desired units, it may be necessary to convert some known parameter or variable from one set of units to another. To do this, *conversion factors* are used. Their intent and utility are self-evident in Illustrative Example 3–2 to follow.

Table 3–2 gives some common conversion factors for the exemplary dimensions and units given in Table 3–1. A more comprehensive list can be found on the Internet (searching for "conversion of English to metric units").

Illustrative Example 3–2: Using Conversion Factors

Newton's second law of motion is given by:

$$F = ma$$

TABLE 3–2　Examples of Conversion Factors for the Dimensions and Units (from Table 3–1)

Metric/SI Unit	English Engineering Unit	Conversion Factor
second (s)	second (sec)	[Identical]
meter (m)	foot (ft)	1 m = 3.28 ft
	inch (in)	1 m = 39.38 in
centimeter (cm)*	inch (in)	1 in = 2.54 cm
kilogram (kg)	pound mass (lbm)	1 kg = 2.205 lbm
gram (g)*	pound mass (lbm)	1 lbm = 453.6 g
Celsius (°C)	Fahrenheit (°F)	T (°C) = 5/9 [T (°F) – 32]
		T (°F) = 9/5 T (°C) + 32
		T (K) = T (°C) + 273
		T (°R) = T (°F) + 459.67
newton (N)	pound force (lbf)	1 lbf = 4.448 N
		1 dyne* = 10^{-5} N
joule (J)	British thermal unit (Btu)	1 Btu = 1054 J
		1 erg* = 10^{-7} J
	calorie* (cal)	1 J = 0.239 cal
		1 Btu = 252.0 cal

*These units are used in the cgs system.

in which F corresponds to the force required to act on (or acting upon) an object of mass m to cause an acceleration a.

If measured values of mass and acceleration are in English units of pounds (lbm) and feet per second per second (ft/s^2), respectively, when force F is needed in metric units of newtons (N = kg-m/s^2), conversion factors are needed for both m and a to generate metric units for each. Appropriate conversion factors to do this are:

$$F \text{ (in N)} = [m \text{ (lbm)}(1 \text{ kg}/2.02 \text{ lbm})][a \text{ (ft/s}^2)(0.3048 \text{ m/ft})]$$

Note that "lbm" cancel between numerator and denominator within the first bracketed expression for m and that "ft" cancel between numerator and denominator in the second bracketed expression for a, leaving units of kg-m/s^2 = N.

In Chapter 5, the importance and value of reconciling units *before* manipulating mathematical equations is discussed.

To deal with the scale of a particular unit (cyclic frequency, stress, length, etc.), prefixes are used with metric SI units. Table 3–3 lists the most important prefixes for SI units, giving the prefix, the value the prefix represents, the standard form, and the symbol used.

Significant Figures

The *significant figures* (also known as *significant digits*) of a number or quantifiable parameter are those digits that actually carry the degree of precision to (i.e., give realistic meaning to) a solution. In other words, they signify the maximum precision one can attribute to a solved-for parameter given the precision with which known parameters in a pertinent mathematical equation are known.

The rules for identifying the number of significant figures when solving a mathematical problem are as follows:

- Nonzero digits are considered significant (e.g., 143 has three significant digits, 1, 4, and 3; 11.45 has four significant digits, 1,1, 4, and 5).
- Zeros that appear anywhere between two nonzero digits are significant (e.g., 104.03 has five significant digits, 1, 0, 4, 0, and 3).
- Leading zeros are not significant (e.g., 0.00143 has only three significant digits, 1, 4, and 3).
- Trailing zeros in a number before or after a decimal point are significant (e.g., 12,400 has five significant digits, 1, 2, 4, 0, and 0; 0.0010400 has five significant digits, 1, 0, 4, 0, and 0). One caveat with trailing zeros is that the trailing zeros are only significant if they are real in the measurement and not simply space fillers (see discussion of "rounding," following Illustrative Example 3–3).

TABLE 3-3 Prefixes for SI Metric Units			
Prefixes	Value	Standard Form	Symbol
Tera	1 000 000 000 000	10^{12}	T
Giga	1 000 000 000	10^{9}	G
Mega	1 000 000	10^{6}	M
Kilo	1 000	10^{3}	k
deci	0.1	10^{-1}	d
centi	0.01	10^{-2}	c
milli	0.001	10^{-3}	m
micro	0.000 001	10^{-6}	μ
nano	0.000 000 001	10^{-9}	n
pico	0.000 000 000 001	10^{-12}	p

The same general rules apply to a number expressed in scientific notation (e.g., 0.025 has two significant digits, 2 and 5, and becomes 2.5×10^{-2}; 0.0003400 has four significant digits and becomes 3.400×10^{-4}).

Illustrative Example 3-3: Significant Figures

If the mass m of an object is measured to be 2.03 kg, and a force F measured to be 10 N ($= kg\text{-}m/s^2$) is applied, the calculated value (using a scientific calculator) for the object's acceleration a from Newton's second law of motion, $F = ma$, rearranged to solve for a is:

$$a = F/m = 10 \text{ N}/2.03 \text{ kg} = 4.926108374 \text{ m/s}^2$$

However, the number of significant digits to which the calculated acceleration a can be known is two, as limited by the precision with which the less precisely measured force F (at 10 N) is known. Thus,

$$a = 4.9 \text{ m/s}^2$$

When significant figures are employed—as they should *always* be—it is often appropriate to round off values for parameters, in other words, to use *rounding*. To round a given or calculated number to n significant digits:

1. Start with the leftmost nonzero digit (e.g., the "2" in 2200 or the "3" in 0.0314).
2. Keep n digits, replace the rest with zeros before a decimal point, and drop all trailing zeros after the last nonzero digit following a decimal point (e.g., for $n = 2$, 4227 becomes 4200 and 0.0213 becomes 0.021; if rounding is appropriate, 4289 becomes 4300 and 0.0218 becomes 0.022).

Rounding is particularly useful when approximating or estimating (see Chapter 4).

Physical and Mathematical Constants

Some mathematical equations used in engineering and in science involve physical or mathematical constants. A *physical constant* is a measured or measurable physical quantity that is generally believed to be universal in nature and constant over time.[4] A *mathematical constant* is likewise fixed in nature and over time but is not—and cannot be—measured.

Examples of some particularly important physical and mathematical constants are given in Table 3–4.

Illustrative Example 3-4: Physical Constants

Einstein's famous equation that expresses the equivalency of mass and energy (i.e., mass-energy equivalence), $E = mc^2$, makes clear why atomic bombs are so devastatingly powerful. A key quantity in the equation is the physical constant c, i.e., the speed of light in a vacuum, which has a value of 3.0×10^8 m/s. The impact of converting 1 kg of fissionable U^{235} totally into energy is seen from:

$$E = (1 \text{ kg})(3.0 \times 10^8 \text{ m/s})^2 = 9 \times 10^{16} \text{ kg-m}^2/s^2 = 9 \times 10^{16} \text{ J}$$

Converted to calories (where 1 calorie represents the energy required to raise 1 g of water by 1°C or 1 K), this is equivalent to:

$$(9 \times 10^{16} \text{ J})(1 \text{ cal}/4.184 \text{ J}) = \sim 2 \times 10^{16} \text{ cal}$$

---

[4]While commonly referred to as a physical constant, g, the acceleration due to gravity, is only constant on the Earth and is *not* "universal." The acceleration due to gravity differs with the mass of an astronomical body, being greater for bodies more massive than Earth and smaller for bodies less massive than Earth.

TABLE 3-4 Examples of Important Physical and Mathematical Constants

Most Important Mathematical Constants*

• Archimedes' constant pi (π)	Circle circumference/diameter	3.14159
• Euler's constant e	Exponential; growth	2.71828
• Conversion of ln to log		2.303†

*There are other mathematical constants, but their use is much more limited and highly specialized; mostly to advanced mathematics.
†This is actually a conversion factor.

Most Important Physical Constants (alphabetically)

• Atomic mass unit, amu	1.66×10^{-27} kg
• Avogadro's number N_A	6.023×10^{23} atoms/molecules per mol
• Bohr magneton μ_B	9.27×10^{-24} J/T
• Bohr radius a_O	5.29×10^{-11} m
• Boltzmann's constant k	1.38×10^{-23} J/K
• Electron rest mass m_e	9.11×10^{-31} kg
• Elementary (electron) charge e	1.602×10^{-19} C (coulombs)
• Faraday constant F	9.65×10^{4} C (coulombs)
• Gas constant R (=kN_A)	8.31 J/mol-K
• Gravitational acceleration g	9.80665 m/s^2 (on Earth)
• Molar volume V_{mol}	22.41383 m^3/kmol
• Neutron rest mass m_n	1.675×10^{-27} kg
• Permeability of vacuum μ_o	1.257×10^{-6} henry/m
• Permittivity of a vacuum ε_o	8.85×10^{-12} farad/m
• Planck's constant h	6.63×10^{-34} J-s
• Proton rest mass m_p	1.67×10^{-27} kg
• Speed of light (in vacuum) c	2.9979×10^{8} m/s

This is enough heat energy to raise the temperature of 2×10^{10} kg (2×10^7 metric tons) of a mass with similar heat capacity to water 1000°C—enough to vaporize, melt, or burn a city[5]! Note that significant figures have been applied to all calculated values.

Equation Forms (The Form of an Equation)

It is often useful to recognize and/or consider the form that a mathematical equation takes, also known as the *form of an equation* in mathematics. As shown in Table 3–5, there are many types or forms of mathematical equations. While the table is not complete, it certainly represents the most commonly used types of equations in engineering.[6]

One value of recognizing the form of a particular mathematical equation in engineering is that it helps the engineer to better understand the physical significance of the equation. For example, for a linear equation with the form $y = mx + b$, m represents the slope of the straight line of a plot of x versus y and b represents the intercept of the line on the y axis where $x = 0$ (i.e., the y intercept). An illustrative example should help.

[5]A *megaton* is used as a measure of the amount of energy released by a nuclear device. It is based on the amount of energy released by the detonation of 1 ton of TNT (trinitrotoluene), which is 4.184×10^9 J, or 4.184 gigajoules (GJ). Given this, 1 megaton is equivalent to 10^{15} J, or 10 petajoules! Fifty-megaton nuclear devices have been tested by both the United States and Russia.
[6]For some of the types or forms of equations given in Table 3–5, there are subtypes. A few examples are included.

TABLE 3-5 Types of Forms of Major Mathematical Equations Used in Engineering

Linear equations	y or x = constant	generate vertical or horizontal lines
	$y = mx + b$	slope-intercept (std.)
	$(y-y_1) = m\,(x-x_1)$	point-slope
	$(y-y_1) = [(y_2-y_1)/(x_2-x_1)](x-x_1)$	two-point
Quadratic equations	$y = ax^2 + bx + c$	generate parabolas or hyperbolas
Cubic equations	$y = ax^3 + bx^2 + cx + d$	
Proportional equations	$y = kx$ or $y = k/x$	to ratio parameters
Equation of a circle	$(x-h)^2 + (y-k)^2 = r^2$	
Equation of an ellipse	$x^2/a^2 + y^2/b^2 = 1$	horizontal axis
	$x^2/b^2 + y^2/a^2 = 1$	vertical axis
Logarithmic equations	$y = \log_b x$	involve logs base b
Exponential equations	$y = bx$, for base b	
Arrhenius equations	$k = A \exp(-E_a/RT)$ per mol	
	or $(-E_a/kT)$ per atom	
Trigonometric equations		involve trig. functions
Ordinary differential equations	$F(x,t) = m\,[d^2x(t)/dt^2]$	Newton's 2nd law

Illustrative Example 3-5: Physical Significance of an Equation

Most metals and alloys (solid solutions of two or more metals) used in engineering applications consist of very large numbers of small, abutting, space-filling, tightly attached crystals called *grains* (i.e., the metals or alloys are *polycrystalline*).[7] It turns out that the strength of a metal or alloy given by the stress level at which it would permanently or plastically deform (i.e., its yield strength σ_y) is given by the Hall-Petch equation, $\sigma_y = \sigma_o + kd^{-1/2}$, in which d is the effective grain diameter (as a representative linear measure of the grain size). The equation indicates that the yield strength of a metal or alloy is increased as the grain size is decreased.[8] In fact, the effect is linear if one plots σ_y versus $d^{-1/2}$, as in Figure 3-1.

The plot in Figure 3-1 makes it apparent that $\sigma_y = \sigma_o + kd^{-1/2}$ has the form of the standard equation for a linear function, $y = b + mx$ (from rearranging $y = mx + b$) in which $y = \sigma_y$, $b = \sigma_o$ = the y intercept, $x = d^{-1/2}$, and $m = k$ = the slope of the line. Thus, in terms of the physical significance of the equation, the slope k relates how fast yield strength changes with a change in grain size and σ_o represents the yield strength of an infinitely large single grain.[9]

---

[7] The size of grains is measured by an ASTM standard (N) that considers the number of grains (n) seen in a 1-in × 1-in (2.54-cm × 2.54-cm) square image at 100X magnification using the relationship that $n = 2^{N-1}$. The number n typically seen ranges from around 32 to 256.

[8] It has been found that as grain size becomes very small, i.e., reaches the nano-scale, the Hall-Petch equation is not followed. Interested readers are encouraged to find out why, using information available on the Internet (e.g., search "Hall-Petch behavior at very small grain sizes").

[9] In fact, it is the presence of boundaries between grains (i.e., grain boundaries) that gives rise to increased resistance to yielding (i.e., higher yield strength) due to the effect the strain energy that exists in the region of distorted crystal planes at each boundary has on crystal imperfections known as *dislocations,* which must move in order for yielding to take place. Hence, more grain boundaries due to smaller grains gives rise to greater yield strength. Correspondingly, the lowest yield strength would occur when there are no grain boundaries, as with a single (infinite) grain.

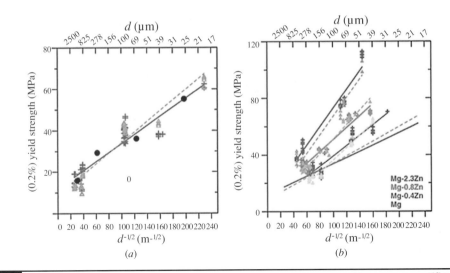

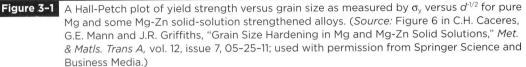

Figure 3-1 A Hall-Petch plot of yield strength versus grain size as measured by σ_y versus $d^{-1/2}$ for pure Mg and some Mg-Zn solid-solution strengthened alloys. (*Source:* Figure 6 in C.H. Caceres, G.E. Mann and J.R. Griffiths, "Grain Size Hardening in Mg and Mg-Zn Solid Solutions," *Met. & Matls. Trans A,* vol. 12, issue 7, 05–25–11; used with permission from Springer Science and Business Media.)

Another very valuable result of recognizing the form of a particular mathematical equation for an engineering property, phenomenon, or process is that it allows the student of engineering (at any point in his or her learning) to see parallels between seemingly unrelated properties, phenomena, or processes. An illustrative example will help.

Illustrative Example 3–6: Recognizing Related Things

At three different times, in three different courses (e.g., heat transfer, basic materials, and physics or circuits theory), from three different teachers, every engineering student learns three equations, as follows:

- Fourier's law of heat transfer: $Q = k(dT/dx)$
- Fick's first law of diffusion: $J = -D(dc/dx)$
- Ohm's law for current: $I = E/R$

While these may seem to be different equations, a little reflection (helped by some restatement for one equation) shows they are all fundamentally the same.

Each equation indicates that how much of some *thing* (on the left side of the equal sign) moves (i.e., its flux) is directly related to how easy it is for that *thing* to move (i.e., its mobility) times how hard it is pushed (i.e., the driving force as imposed by a gradient, on the right side of the equal sign).

To see this for electrical current, Ohm's law needs to be restated using the relationship that the resistance R across a physical material object is the product of the inherent hindrance a material fundamentally presents to the movement of current (as a flow of electrons), known as resistivity ρ, and the ratio of a physical material object's length L and cross-sectional area A in $R = \rho(L/A)$, recognizing that the electrical conductivity σ_{el} and the electrical resistivity ρ have a reciprocal relationship, i.e., $\sigma_{el} = 1/\rho$, and that E (the electric potential) is really a gradient in voltage over a distance (i.e., dV/dx). Thus, Ohm's law restated, with some liberty to ignore the effect of the physical size of the conductor, can be written as follows.

$$I = \sigma_{el}(dV/dx)$$

Hence, all three equations have the same mathematical form and, thus, denote similar physical behavior; that is, the flux of heat (Q in Fourier's equation), flux of mass (J in Fick's equation), and flux of charge carriers (I in Ohm's restated equation) all depend on the product of mobility of the heat (i.e., given by thermal conductivity k), mass (i.e., given by diffusion coefficient or diffusivity D), and charge (i.e., given by electrical conductivity σ_{el}), and the gradients of temperature (dT/dx), concentration (dc/dx), and voltage (dV/dx), respectively, as driving forces.

For the author, this makes sense and makes understanding how Nature works much simpler. There are only three fundamental *things* that can (or need to) move in Nature: heat, mass, or charge. The movement (i.e., flux) of all three things is controlled by the same factors (i.e., mobility and driving force in the form of a gradient), as a product. Making engineering simpler to understand by students and practitioners is a good thing and does not detract from what engineers do. It only improves understanding.

There are many, many examples in engineering where similar forms of governing mathematical equations for seeming unrelated properties, phenomena, or processes indicate a fundamental basis.[10]

Pulling It All Together

A final illustrative example that is somewhat more elaborate should help to pull together what has been covered in this chapter: (1) the use of mathematical equations to solve closed-ended quantitative problems; the importance of (2) units, (3) conversion factors, and (4) physical and/or mathematical constants in those equations; and (5) the value (if not the importance) of recognizing and considering the form of the mathematical equation(s).

Illustrative Example 3–7: Pulling It All Together

The atoms in crystalline metals and alloys move around (i.e., migrate) within the solid by an atomic-level process known as *diffusion,* and they do so *down* any prevailing concentration gradient for the diffusing species (see Fick's first law in Illustrative Example 3–6), which involves dc/dx. The rate at which they do this is given by their diffusion coefficient or diffusivity D (with units of m^2/s or cm^2/s)[11] to reflect flow across an imaginary plane of unit area (either m^2 or cm^2) per unit of time (s), which follows temperature dependency expressed by the mathematical expression that[12]:

$$D = D_o \exp(-Q_d/RT)$$

In this equation, temperature T has units of absolute degrees Kelvin (K), R is the physical gas constant (= 8.31 J/mol-k in the kgs system or 1.987 cal/mol-K in the cgs system), Q_d is the activation energy barrier that 1 mole of atoms must overcome to move by atomic jumps, in other words, to diffuse, with units of J/mol or cal/mol, and D_o is a preexponential

---

[10] An important example is that the temperature dependence of D on T in the exponential relationship (in Illustrative Example 3–7) is the direct result of the exponential relationships between the likelihood of successful jumps by a diffusing atom and temperature.

[11] For each parameter, kgs units, commonly used in engineering, are given first, and cgs units, commonly used in science, are given second.

[12] An equation of this form is exponential in terms of T, and is commonly referred to as an *Arrhenius equation.*

temperature-independent constant (but not a physical or mathematical constant), also with units m^2/s or cm^2/s.

An engineer is faced with having to determine the values of the activation energy for diffusion Q_d and the preexponential diffusion constant D_o for some alloy for which, let's presume, there are no data available. To do this, he or she decides to experimentally measure the rate of diffusion D of the diffusing species B in the host metal A at two different temperatures, T_1 low and T_2 high. This is, in fact, how values of Q_d and D_o are found, recalling (from a basic materials course in college) that D is temperature dependent by $D = D_o \exp(-Q_d/RT)$.

A spark of recognition flashes in the engineer's brain: This equation can be plotted on a semilogarithmic plot of D versus reciprocal temperature $1/T$ to yield a straight-line behavior. So the engineer calculates the natural logarithm of the equation and gets:

$$\ln D = \ln D_o - Q_d/RT$$

which can be rearranged to more clearly show the relationship between D and T (as $1/T$) by:

$$\ln D = \ln D_o - (Q_d/R)(1/T)$$

Furthermore, the engineer thinks it might be useful to express this new, rearranged equation in terms of base-10 logarithms, giving:

$$\log D = \log D_o - (Q_d/2.3R)(1/T)$$

(The number 2.3 is a really not a mathematical constant but, rather, a conversion factor that relates base-e and base-10 logarithms.)

Since D_o, Q_d, and R are all constants, this equation is seen to have the form of the standard equation for a straight line, that is (slightly rearranged from normal):

$$y = b + mx$$

in which y and x are analogous, respectively, to the variables $\log D$ and $1/T$ in the preceding equation.

Thus, the engineer decides to plot $\log D$ versus $1/T$ for the two measured values he or she obtained experimentally, giving a plot like that shown in Figure 3-2.

Looking at this plot, it becomes clear that the slope m of the line has a value corresponding to $-Q_d/2.3R$, which will allow Q_d to be determined, by rearranging this term, as:

$$Q_d = -2.3R\left[(\log D_1 - \log D_2)/(1/T_1 - 1/T_2)\right]$$

in which the bracketed expression gives the slope from the change in D (as $\log D$) for a change in T (as $1/T$).

If, for further clarity, values of $1/T_1$ and $1/T_2$ of 0.80 and 1.1 in units of 1000/K (corresponding to $T_1 = 1250$ K and $T_2 = 900$ K) were used in the experiment, and the log values for the diffusivity measured at each temperature were $D_{900} = -12.40$ and $D_{1250} = -15.45$, the value of Q_d can be calculated from:

$$2.3\,(8.31\text{ J/mol-K})\{[(-12.40)-(-15.45)]/[0.8 \times 10^{-3}\text{ K}^{-1}-1.1 \times 10^{-3}\text{ K}^{-1}]\}$$

to give $Q_d = 194{,}000$ J/mol, or 194 kJ/mol.

Using the rules given earlier, the calculated value for Q_d is accurate to three significant digits because 2.3 is actually precise to only three significant digits if fully expressed as 2.30, and the reciprocal temperature values of 0.8×10^{-3} and 1.1×10^{-3} actually represent temperatures (used in the experiment) presumed to be 1250 and 900 K, respectively.

Rather than attempt to obtain a value for D_o by extrapolating the straight-line plot to intercept the y axis (made difficult by a reciprocal scale for temperature in which $1/T$ would

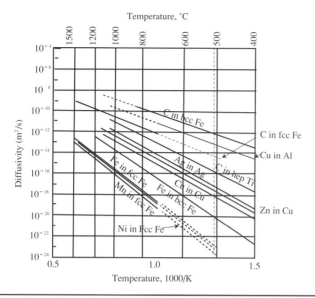

Temperature, °C

Diffusivity (m^2/s)

Temperature, 1000/K

Figure 3-2 Plot of the logarithm (base-10) of the diffusion rate (or diffusivity) D (which has units of m^2/s) versus Reciprocal Temperature (1/T) in units of 1000/K for a diffusing species (i.e., solute) in a host (or solvent) for which values of Q_d and D_o were unavailable. (*Source:* Data extracted from several references.)

have a value of zero on the x axis only if T was infinite), it is easier—and more prudent—to calculate a value for either measured value of D—say, D_{800} or D_{1250}—using the rearranged equation

$$\log D_o = \log D + (Q_d/2.3R)(1/T)$$

In this equation, if T = 900 K, for which the value of D_{900} would have been 3.5×10^{-16} m^2/s and for which the log D_{900} is –15.45, the result for log D_o would be obtained from:

$$-15.45 + \{[(194,000 \text{ J/mol})/(2.3)(8.31 \text{ J/mol-K})][1.1 \times 10^{-3} \text{ K}^{-1}]$$

to give:

$$\log D_o = -4.28$$

which corresponds to a value for D_o = 5.2×10^{-5} m^2/s.

If, for some reason, it were necessary to convert Q_d and D_o from kgs to cgs units, this could be done using appropriate conversion factors, as in the following:

$$Q_d = 194,000 \text{ J/mol-K } (1 \text{ cal}/4.184 \text{ J}) = 46,400 \text{ cal/mol-K}$$

and

$$D_o = 5.2 \times 10^{-5} \text{ } m^2\text{/s } (10^4 \text{ cm}^2/m^2) = 5.2 \times 10^{-1} \text{ cm}^2\text{/s}$$

In Summary

This chapter describes how mathematical equations may be used to solve problems for which there is a single, quantifiable analytical solution. When using mathematical equations, one should: (1) insert units for all parameters, known and unknown; (2) use appropriate conversion factors to reconcile units for the known parameters to generate the solution in the needed or desired units; (3) employ the rules for significant figures to reflect that a quantitative solution can be no more precise than the least precise input parameter in an equation; (4) employ physical and/or mathematical constants where called for in an equation; and (5) be aware of, recognize, and consider the form of the mathematical equation to (a) make clear the physical significance of equation and/or (b) allow parallels to be recognized between seemingly unrelated properties, phenomena, or processes.

Suggested Resources

httl://www.en.wikipedia.org/wiki/Trigonometric_functions

http://www.ligo.caltech.edu/~vsanni/ph3/SignificantFiguresAnd Measurements/
   SignificantFiguresAndMeasurements.pdf

http://www.themathpage.com, search under "Topics in Precalculus."

http://www.wikipedia.com, search under "Algebraic Functions" and "Types of Equations."

CHAPTER 4

Approximating and Estimating

To MANY PEOPLE, *approximating* something and *estimating* something suggest the same thing. But, for engineers—and in engineering—there are important differences between these two as problem-solving techniques. The differences become apparent, in fact, if one checks the definitions of *approximate* and *estimate* (the root words in the gerund—"-ing"—forms, which, like all gerunds tend to function like noun-verbs, i.e., things having action).[1]

According to the dictionary, *approximate,* as a noun, means "almost exact or correct; very similar; closely resembling." For problem-solving, the noun becomes an action. In contrast, *estimate,* as a verb, means "to calculate approximately the amount, extent, position, or value of something." Both *approximating* and *estimating* are valuable techniques in engineering problem-solving. Some help, but let's delve a little more deeply.

Approximating

It is sometimes the goal of an engineer when approaching a problem that could be solved precisely using an equation or equations (i.e., analytically) to only get an idea of what the complete and precise solution would be. *Approximating* should be considered for use if (1) the time to completely and precisely solve an equation or set of equations is not available (e.g., one is on the run), (2) it would not be prudent to spend the time required to completely and precisely solve the equation(s) because of the time (and, in practice, cost) it would involve for a complete solution (which may not be merited before a job is fully undertaken, for example), or (3) because the resources needed to obtain a complete and precise solution are not at hand (e.g., one has no electronic calculator, laptop, or PC).

Approximating very much fits with the author's philosophy in engineering to "never measure a manure pile with a micrometer." (After all, it's just a pile of poop!) Many times, all an engineer really needs is to get a sense of something, not a detailed solution and/or a precise number.

It is not uncommon for practicing engineers to talk about "back-of-the-envelope" solutions. What they generally mean is that they will run a quick approximate solution on a handy piece of paper—often in the absence of any electronic assist (e.g., calculator or computer). Sometimes engineers—especially experienced ones—approximate a number in their head.

How practical or easy it will be to approximate a solution to an equation obviously depends on the form of the equation involved (see Chapter 3). It is practical and usually easy to do for linear algebraic equations. It is tougher to do for trigonometric equations, although some engineers can do it using

---

[1]Definitions are from Houghton Mifflin's online dictionary at www.thefreedictionary.com/Houghton.

their basic knowledge of a sine curve.[2] For most engineers, approximating logarithmic or exponential equations is too difficult to do, unless they use them all the time.[3]

There is another underrated—at least, underemphasized—value of approximating each and every time a mathematical approach using equations is employed. An approximate solution helps an engineer (and especially a student!) know whether a precise solution, once obtained, is correct, by fixing the relative magnitude of the solution. Approximating used this way is a method of checking the mathematics to ensure that careless, unnecessary, and, for engineers, unacceptable mistakes were not made.[4] The use of approximating for this purpose is especially helpful when values of parameters are expressed in scientific notation (i.e., powers of 10).

An illustrative example of approximating to "size" an answer *and* to check the mathematics should help.

Illustrative Example 4-1: Approximating an Answer/Checking Math

The electrical conductivity σ_{el} of an intrinsic (nondoped) semiconductor is given by the equation:

$$\sigma_{el} = n_i \, |e| \, (\mu_e + \mu_h)$$

in which n_i is the number of charge carriers per unit volume in the semiconductor (here, for intrinsic gallium arsenide, GaAs, taken to be 7.0×10^{12} m^{-3}), $|e|$ is the absolute value of the physical constant for the charge on an electron (i.e., $e = 1.602 \times 10^{-19}$ C, where a coulomb [C] is an amp-second, A-s), and μ_e and μ_h are the mobility of electrons (as negative charge carriers) and holes (being the absence of electrons from where they ought to be, as positive charge carriers), here, for GaAs, having values of 0.85 and 0.04 m^2/V-s, respectively.

Inserting these values will allow a precise value of σ_{el} to be calculated from:

$$\sigma_{el} = (7.0 \times 10^{12} \text{ m}^{-3})(1.602 \times 10^{-19} \text{ A-s})(0.85 + 0.04 \text{ m}^2/\text{V-s})$$

Before doing this, approximating the answer using rounding gives:

$$\sigma_{el} \sim (7 \times 10^{12})(1.6 \times 10^{-19})(0.9) \sim (11.2 \times 0.9) \times 10^{-7} \sim 10 \times 10^{-7} \sim 10^{-6}$$

Reconciling units gives:

$$(\text{m}^{-3})(\text{A-s})(\text{m}^2/\text{V-s}) \text{ or } \Omega^{-1}\text{-m}^{-1} \text{ or } (\Omega\text{-m})^{-1}$$

given that an ohm (Ω) is V/A.
So the expected answer should be approximately 10^{-6} $(\Omega\text{-m})^{-1}$.
Precise arithmetic solution gives:

$$\sigma_{el} = 9.98 \times 10^{-7} \text{ } (\Omega\text{-m})^{-1}$$

which, to two significant figures, is 10×10^{-7} or 10^{-6} $(\Omega\text{-m})^{-1}$. The estimate is remarkably—and fortuitously—close.

[2]One should know, as an engineer, that the sine function has values of 0 at 0°, 180°, and 360°; +1 at 90° and −1 at 270°; +1/2 at 30° and 150° and −1/2 at 210° and 330°; and +0.707 at 45° and 135° and −0.707 at 225° and 315°. Picture the sine curve plot. The cosine function is shifted 90° from the sine function, with cos 0° = 1 and cos 90° = 0.

[3]One handy fact for approximating with an exponential relationship is that for an Arrhenius equation such as for the rate of diffusion D, in which $D = D_o \exp(-Q_d/RT)$. The rate of diffusion doubles for each 30°C/50°F increase in temperature T or is halved for each 30°C/50°F decrease. This fact is useful for approximating the effect of a change in the heat treatment temperature (e.g., to temper a quench hardened steel or age a solution treated Al alloy) on the time to achieve equivalent effects at different temperatures.

[4]Checking, as a problem-solving technique, is discussed in Chapter 6.

TABLE 4-1	Summary of the Differences Between Approximating and Estimating as Engineering Problem-Solving Techniques	
Approximating	**Estimating**
• For roughly sizing a number	• For assessing feasibility of an idea
• For getting a number quickly	• For calculating material, labor, or cost during planning
• For checking math via comparison	
Involves "back-of-the-envelope"	*Involves "rough-order-of-magnitude" initially, and greater detail and accuracy later*

Obviously, if an error in math had been made anywhere during the solution of the equation, it would be made clear by comparing the full solution to the approximate solution. One of the most common errors occurs when operating with + and/or − exponentials in both numerators and denominators. Checking while approximating helps catch such errors.

Estimating

According to the dictionary definition, *estimating* involves calculation of an approximate number for something. This is where estimating and approximating are intermingled and, as a result, the difference between the two becomes confusing. In this book, *estimating* should be considered as a potential problem-solving technique in either of two situations:

1. To assess the *feasibility* of a particular engineering idea prior to embarking on a design effort on either (a) a technological basis or (b) an economic (or cost) basis
2. As a planning tool in design for material, labor, and/or cost, and so forth, approximately at first and in more detail and with greater precision later, when the estimate approaches the real number

A concise summary of the differences between approximating and estimating is given in Table 4–1.

Sometimes ideas emerge for a new product, for example, which may not have been thought through, whether by a client or by the designer/manufacturer. For such ideas, not only would it be economically wasteful to proceed with all of the problem-solving effort (and cost) that is associated with design, but it would prove to be an embarrassment, as it would suggest technical and/or economic naiveté, at best, and ignorance, at worst. Neither does anything to further an engineer's career. If there is any question of the feasibility, viability, or practicality of an idea, estimating should be used as a determinant. An illustrative example will help.

Illustrative Example 4-2: Estimating Feasibility

A well-meaning entrepreneur, a philanthropist concerned about CO_2 emissions to the atmosphere as a possible contributor to global warming,[5] approached the design center at the author's university a decade ago with an idea he wished to fund. He proposed addressing two problems at once: (1) to extract energy from the wind (for conversion

[5]If global warming is actually taking place—which must be determined with objective data and free of emotion—it must next be determined from what source (a weakening magnetic field surrounding Earth, "greenhouse effect" from CO_2 from what source, natural or man-made, etc.). Remember, to solve a problem one needs to know if there is a problem and precisely what that problem is!

to electricity) using pollution-free Savonius-type vertical-axis wind turbines[6] and (2) to remove some portion of the excess (i.e., man-made) CO_2 from the atmosphere using a "coating of a CO_2-absorbing alkali-metal hydroxide in water" applied to the turbine's blade surfaces. The CO_2 would react with the alkali-metal hydroxide to produce a carbonate of the alkali metal, which could be "removed from solution" and "used for some manufacturing process."[7] (The text in quotation marks represents the entrepreneur's words.)

Out of youthful enthusiasm and a strong desire to do something positive for the environment, more than 30 seniors engaged in a capstone design course "jumped into the pool" [the author's words] to generate alternative design concepts from the problem definition. The author, as a seasoned engineer with industrial experience, and serving as a faculty advisor to a couple of the half-dozen teams of students, challenged the feasibility of such an idea. The students mistook the technical challenge as a lack of concern for the environment, which it was not. The objective was simply to teach these engineers-in-training to think before they act. They had been acting for more than two weeks, albeit, without thinking first! To convince them, the author—as their faculty advisor—generated a crude estimate of feasibility in two hours over a weekend.

The thought process used to estimate the feasibility of the idea involved finding out how much CO_2 could be removed by such an approach per wind turbine, to get an idea of how many wind turbines would be required to have some reasonably favorable impact. Some of the costs involved would also be estimated. The steps used in the process were:

1. The magnitude of the CO_2 emission problem was found from data available online (e.g., www.en.wikipedia.org) under a search for "CO_2 emissions by country") here, for 2007, the latest year available:

    29,321,302 metric tons/year for the world
    5,833,381 metric tons for the United States (~19.9 percent)[8]

    Since balanced chemical reactions will be involved, the approximate number of moles of new CO_2 emissions in the world, per year, was calculated using a rounded value for the world in 2007, to two significant figures, as:

    $$(29 \times 10^6 \text{ metric tons})(10^3 \text{ kg/ton})(10^3 \text{ g/kg})(1 \text{ mole of } CO_2/44 \text{ g}) =$$
    $$6.6 \times 10^{11} \text{ moles of } CO_2 \text{ per year}$$

2. The balanced operative chemical reaction between CO_2 and the most effective and readily available alkali-metal hydroxide, KOH,[9] is:

    $$2KOH + CO_2 \Rightarrow K_2CO_3 + H_2O$$

3. A half-shell cylindrical Savonius vertical-axis wind turbine design was chosen for its simple geometry for easy estimating and construction.

    As an approximation, it was assumed that a complete turbine system would consist of three stacked units (as shown in Figure 4–1a), in which each unit (for easy math, to give an approximate answer) would be 4 m in diameter D (as shown in

---

[6]The advantage of vertical-axis wind turbines is that they capture the wind from any direction without having to turn into the wind as horizontal-axis bladed turbines do. The disadvantage of the Savonius types is that they are rather inefficient for electrical energy generation.

[7]The entrepreneur erroneously, it turns out, assumed the carbonate would be insoluble in water and, so, would precipitate out as a solid that could be removed by filtering. Not all alkali-metal carbonates are insoluble in water.

[8]Incidentally, the suggested accuracy of these numbers—to 7 or 8 significant digits—is absurd. No one could calculate—or likely measure—CO_2 emissions to this degree of precision. It causes—or should cause—one to question the source of the data. Maybe 5.9 million metric tons for the United States would do.

[9]In order, from most effective to least effective, these are LiOH, KOH, and NaOH. However, LiOH is far less common and far more expensive than KOH.

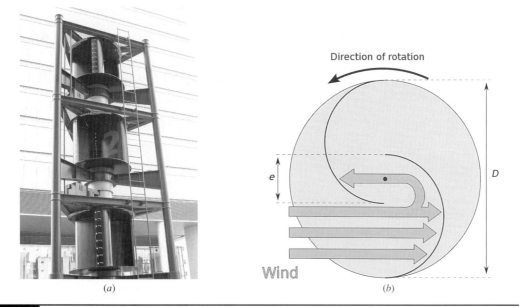

(a)

(b)

Direction of rotation

e

D

Wind

Figure 4-1 A Savonius vertical-axis wind turbine of a half-shell design stacked three high (*a*), showing geometric details (*b*). (*Source:* www.en.wikipedia.wikipedia, from a Google image search; [*a*] originally contributed by Yuval Y on 3 December 2007 and [*b*] originally contributed by Ugo14 on 16 February 2006, both used under Creative Commons Attribution 3.0, for which permission is not required.)

Figure 4–1*b*) and 4 m high; so that half-shells would each have a diameter *d* of 2 m and a height of 4 m. This resulted in a total wind-side surface area SA (to which the KOH-solution would be applied) of:

$$SA = (2 \text{ half-shells/unit})(3 \text{ units/system})(\pi dh/2 \text{ m}^2/\text{half-shell})$$
$$= (2)(3)(3.14)(2 \text{ m})(4 \text{ m})/2 = \sim 75 \text{ m}^2/\text{system}$$

4. To work, a saturated solution of KOH in water would have to coat each half-shell's wind-side surface (i.e., estimated at ~75 m² per turbine system) with a film. Such a film could be made to flow continuously over the surface, but would require expensive pumps and an online refresh system to extract K_2CO_3 by some type of exchange precipitation reaction, as K_2CO_3 is *water-soluble*, which could be expensive. For approximating purposes only, it was assumed that a static film of KOH solution 1 mm (i.e., 10^{-1} cm) thick (*t*) would coat the half-shell surfaces, giving a volume KOH per turbine system of:

$$V = (SA)(t) = (75 \text{ m}^2)(10^4 \text{ cm}^2/\text{m}^2)(10^{-1} \text{ cm}) = 7.5 \times 10^4 \text{ cm}^3$$

Water alone (without dissolved KOH, with a density or specific gravity 2.044 times that of water) of this volume would weigh *W*:

$$W = 7.5 \times 10^3 \text{ cm}^3(1 \text{ g/cm}^3) = 7.5 \times 10^3 \text{ g} = 7.5 \text{ kg (16.5 lbm)}$$

This added weight would not be a problem, although this couldn't have been known without estimating the effect.

5. The balanced chemical reaction shows that removing 1 mole of CO_2 will require 2 moles of KOH and produce 1 mole of K_2CO_3, the former of which will have to be bought and the latter of which will have to either be consumed in some manufacturing process or disposed of.

To remove just 1 percent of the 6.6×10^{11} moles of excess CO_2 entering the atmosphere each year (from Step 1), or 6.6×10^9 moles, 13.2×10^9 moles of KOH would be required (and consumed) and 6.6×10^9 moles of K_2CO_3 would be produced (to be dealt with). Converted to weights (from molecular weights of 56.1 and 122.2 g/mole for KOH and K_2CO_3, respectively), these are:

$$\text{Weight of KOH needed} = (13.2 \times 10^9 \text{ moles})(56.1 \text{ g/mole})$$
$$= 7.4 \times 10^{11} \text{ g} = 7.4 \times 10^8 \text{ kg} = 7.4 \times 10^5 \text{ metric tons}$$

$$\text{Weight of } K_2CO_3 \text{ produced} = (6.6 \times 10^9 \text{ moles})(122.2 \text{ g/mole})$$
$$= 8 \times 10^{11} \text{ g} = 8 \times 10^8 \text{ kg} = 8 \times 10^5 \text{ metric tons}$$

While not staggering, neither are these quantities insignificant. A current price for KOH is about $5 per kilogram, so the estimated cost of KOH to remove 1 percent of the excess CO_2 would be:

$$(7.4 \times 10^8 \text{ kg})(\$5/\text{kg}) = \$37 \times 10^8, \text{ or } \$3.7\text{B}$$

That's a lot of money to have a 1 percent impact!

6. If each Savonius wind turbine system (as described here) operated 24 hours a day, 365 days per year, the KOH solution would have to be refreshed at a rate to match the rate of CO_2 absorption and reaction to form K_2CO_3, which would have to be extracted. In order to estimate how much CO_2 each turbine system could remove per day, as a means of estimating how many such turbine systems would be required (around the world) to remove just 1 percent of the excess CO_2, the students were directed to measure how long it took for CO_2 to be absorbed and react.

To do this, a 1-molar (1M) solution of KOH in water at 20°C was sprayed, as a mist, into a chamber containing 100% CO_2, and, after various times, solution was removed to determine (using chemical analysis at a laboratory) how much K_2CO_3 was produced. This allowed the amount of CO_2 absorbed to be determined, as 1 mole of K_2CO_3 is produced for each 1 mole of CO_2 absorbed and reacted.

The experimentally measured time for saturating the 1M KOH solution with CO_2 was over 1 hour; and this was for a fine spray for which the CO_2-absorbing surface area was much greater than for a flowing film of solution.[10]

For this rate of absorption and reaction, the amount of CO_2 that could be removed by a single turbine system in one year could be approximated as follows:

Number of moles of CO_2 absorbed by a film of water alone on the half-shell surfaces, using numbers generated earlier:

$$(7.5 \times 10^4 \text{ cm}^3 \text{ } H_2O)(1 \text{ L}/10^3 \text{ cm}^3)(1.8 \text{ g/L max. solubility } CO_2)[11]$$
$$= 135 \text{ g } CO_2 \text{ per film or } (135 \text{ g})(1 \text{ mole}/44 \text{ g}) = \sim 3.1 \text{ moles}$$

For a film refreshed or replaced every hour, this would allow

$$(3.1 \text{ moles } CO_2/\text{hr})(24 \text{ hr/day})(365 \text{ days/yr})$$
$$= 27{,}200 \text{ moles of } CO_2 \text{ removed per year per system}$$

Thus, to remove 1 percent of the excess CO_2 per year (i.e., 6.6×10^9 moles) would require:

$$6.6 \times 10^9 \text{ moles of } CO_2/(27.2 \times 10^3 \text{ moles/turbine system})$$
$$= 2.4 \times 10^5 \text{ turbine systems}$$

Again, not staggering, but surely nontrivial.

[10]There are ways, of course, that an engineer could deal with this, e.g., by saturating a very high-surface-area felt with KOH solution.

[11]Recall that 1 cm³ of water is equivalent to 1 mL or 10^{-3} L.

While the system might work, the impact—at just 1 percent using 240,000 Savonius wind turbines—is of questionable economic viability. And there are lots of engineering problems to be overcome in dealing with details (distributing the KOH-solution on the half-shell surfaces, refreshing the KOH solution, removal and either use or disposal of the K_2CO_3, etc.). Potassium carbonate (K_2CO_3 or potash) is abundant in nature and is used in the manufacture of soap and glass—but could another 8.5×10^5 metric tonnes be used? The lesson is: Generate approximate numbers and estimate technical feasibility—possibly with economic viability—before engaging in design!

Since engineers make things (which they or other engineers designed), they are frequently faced with the problem(s) of deciding (1) how much input (or raw) material(s) will be required to produce the needed output, (2) how much time will be required to complete a task or set of tasks (i.e., a job), and/or (3) how much something will cost. To do any or all of these, they use estimating.

Estimating may only be to "get an idea," that is, a rough order of magnitude, of what is required or involved, or, in most cases, at some point, it might need to become more precise, the process involved being much more detailed and elaborate. In the latter case, estimating might move toward accounting.

Estimating often involves shortcuts. For repetitive problems (e.g., estimating the cost to install a hardwood floor of Brazilian cherry or estimating the cost to erect a metal building), software packages are commonly used. With these, the engineer, or, in some cases, the prospective customer, enters basic information (such as length and width of a room to be refloored or the base dimensions for a planned metal building, with the cost estimate being generated automatically).[12]

It is not unusual in architecture and/or commercial building construction for estimating to be done using average cost per square foot (or per square meter). Figure 4–2 gives examples for the estimated cost per square foot in U.S. dollars to construct a typical 4- to 8-story hospital (Figure 4–2a) and for a typical 11- to 20-story commercial office building in the United States, by major city (Figure 4–2b). Obviously, engineers familiar with what is involved in the construction of such buildings came up with a "wraparound" cost per unit area for use in the plots.

Estimating can, without question, become very involved as a problem becomes more complex. Just imagine what is involved in order for a construction company to submit a cost quotation as part of a proposal to construct a new highway. Estimates of raw materials (sand, gravel, crushed stone for ballast under the roadway, concrete, steel reinforcing bars and wire mesh, etc.), labor (to survey, clear trees and brush, excavate and/or fill, level and grade, set forms and lay reinforcing bar or wire mesh, pour concrete, etc.), equipment (bulldozers, graders, dump trucks, cranes, etc.), and personnel (cement workers, steelworkers, welders, flag workers, etc.). In addition, every day that construction takes place to widen an existing highway, for example, traffic cones (for which there is a cost) must be placed before work begins and removed when work is stopped for the day. If work takes place at night, the same thing must be done for lights. And, to ensure success, the estimate must be, at one and the same time, low enough to be considered reasonable, credible, and competitive *yet* high enough to allow a profit to be made.

An illustrative example should help.

Illustrative Example 4–3: Material Estimating

The small privately owned company for which you work as the only degreed engineer is expanding its manufacturing capability to include large-capacity injection molding machines for making thermoplastic parts. The owner has sketched plans for a 3200-ft^2

---

[12]An example can be found for metal buildings at http://www.buildingsguide.com/estimates/building-cost-estimate.php.

HOSPITAL (4 TO 8 STORIES) CONSTRUCTION COST:
DECEMBER 2011 RANKING OF MAJOR U.S. CITIES

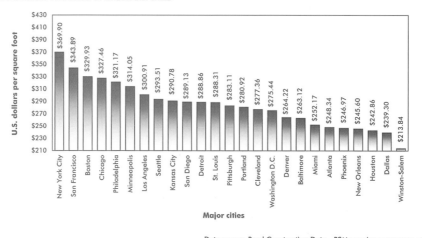

Data source: Reed Construction Data - RSMeans (www.rsmeans.com)
Chart: Reed Construction Data - CanaData.

(*a*)

OFFICE BUILDING (11 TO 20 STORIES) CONSTRUCTION COST:
JANUARY 2012 RANKING OF MAJOR U.S. CITIES

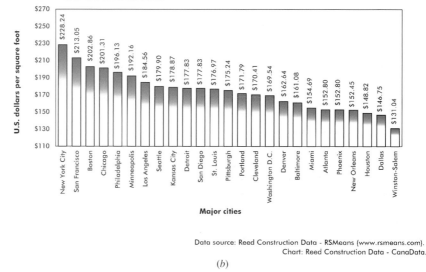

Data source: Reed Construction Data - RSMeans (www.rsmeans.com).
Chart: Reed Construction Data - CanaData.

(*b*)

Figure 4-2 Plots, in the form of bar graphs, showing the average $US cost per square foot to construct a typical 4- to 8-story hospital (*a*) or 11- to 20-story commercial office building (*b*) in various major cities in the United States. (*Source:* Reed Construction Data-Canada, copyright RSMeans, adapted fropm RSMeans Square Foot Costs, used with permission.)

40-ft × 80-ft ($W \times L$) addition to be housed within a commercially available straight-walled metal building. The injection molding machine manufacturer has advised the owner that the machines to be purchased and installed are extremely heavy and so require a reinforced concrete floor with a capacity to safely handle 120- to 125-lbm/ft² loading. To be on the safe side, the owner tells you to plan on 150 lbm/ft². He asks you to estimate the amounts and delivered costs for the needed concrete and steel reinforcing bar, as well as for the metal building. He said he would handle finding a local contractor to do the concrete work.

After a little research, you determine that a concrete pad to handle 150 lbm/ft² must be 9 in thick (*t*), with 1/2-in-diameter (#4) steel reinforcing bar (rebar), spaced on 8-in centers in both directions, at the midthickness of the concrete. You also found details for "control joints" to restrict any cracking of the pad to acceptable areas and "expansion joints" to deal with thermal expansion and contraction.

You obtain the volume of the rectangular pad from the equation:

$$V = W \times L \times t = (40 \text{ ft})(80 \text{ ft})(9 \text{ in})(1 \text{ ft}/12 \text{ in}) = 2400 \text{ ft}^3$$

You convert this volume to cubic yards (cu. yd. or yd³), since that is how concrete is delivered and priced in the United States, giving:

$$2400 \text{ ft}^3 (1 \text{ yd}^3/27 \text{ ft}^3) = 88.89, \text{ which rounds to } 90 \text{ yd}^3$$

The delivered cost for concrete (available on the Internet), as "full loads" of 10 yd³, is $120 to $150/yd³, so you estimate:

$$90 \text{ yd}^3 (\$150/ \text{ yd}^3) = \$13,500 \text{ for concrete}$$

If steel reinforcement is to be placed on 8-in centers, you calculate the total number of linear feet of 1/2-in rebar using the following, in two steps: the first for the width dimension and the second for the length dimension.

Linear feet for the width = (40 ft)(12 in/ft)(1 run/8 in) × (80 ft /run) = 4800 ft
Linear feet for the length = (80 ft)(12 in/ft)(1 run/8 in) × (40 ft/run) = 4800 ft

Thus, 9600 linear feet of #4 rebar is needed, at a delivered cost of $9 per 6-ft length, for a total cost for delivered rebar of:

$$(9,600 \text{ ft})(1 \text{ rebar}/6 \text{ ft})(\$9/\text{rebar}) = \$14,400$$

You are comfortable with your estimate for concrete, as you have an extra cubic yard. You think you probably should order some additional rebar (beyond the 1600 6-ft lengths needed, to account for losses during cutting, bending, and so on, so you suggest 1 percent (16) extra bars, at an additional cost of $144.

To find the cost of a commercial straight-wall metal building, you go to http://metalbuildingsguide.com/estimates/building-cost-estimate.php and obtain the following estimate using the available online estimator for a 40-ft × 80-ft building:

Base building:	$25,600
Accessories:	$5,120
Delivery:	$2,048
	$32,768

A local steel fabricator would have to be hired to erect the delivered building.

Obviously, estimating can be much more involved as the jobs—and problems—get bigger.

In Summary

Approximating in engineering is used of necessity to get an idea of what a complete and precise solution to an analytical problem involving an equation or set of equations will be, but it should always be used to check whether the complete and precise solution obtained is "in the ballpark." *Estimating* is used in engineering during the planning stage of a problem or job. Estimates may only need to be approximate at first but typically need to become more detailed and accurate as one progresses toward having to actually solve a problem. For complex, multitask or multicomponent jobs, estimating must be, at once, close enough to allow a job to be won, without losing money. A much more important role for estimating in engineering, however, is for assessing the technical feasibility or practicality and/or cost viability of a novel idea *before* "jumping into the pool" to begin design.

CHAPTER 5

Interpolating
and Extrapolating

IN MATHEMATICS, the related techniques of *interpolating* and *extrapolating* are defined as follows[1]:

- *Interpolating* is the action of estimating (as used in mathematics to correlate the approximating, as used in Chapter 4) a value for a function <u>between</u> values already known or determined, as data points.
- *Extrapolating* is the action of estimating a value for a function outside [or <u>beyond</u>] values already known, or determined, by assuming that the estimated values follow logically from the known values.

The key words are "between" and "beyond."

In engineering and science, specific values of a needed dependent variable, as data points, are often obtained by calculation or, more often, by limited experiments (for specific conditions) or by sampling outputs during a process (under specific conditions) within a range of an independent parameter or input. While the actual process of interpolation can involve physically constructing new data points (graphically, as covered in Chapter 6) within the range of known data points, it can also be done by mathematically estimating intermediate values. The technique of extrapolation, on the other hand, almost always involves physically extending the curve for a plot of data points within a range, upward or downward, on the axis for the independent parameter (or input), but can also be done mathematically.

Interpolating

There are several different interpolation methods, the most common being:

- Linear interpolation
- Polynomial interpolation
- Spline interpolation

Which method should be used depends on the rate at which the dependent parameter, or output, changes with a change in an independent parameter, or input (see Chapter 12, "Response Curves and Surfaces"), and, obviously, further depends on the accuracy with which the value of an intermediate point is needed. These effects can be seen schematically from Figure 5–1.

Linear interpolation works best for straight-line plots and reasonably well for curves that can be reasonably approximated by straight-line segments between known data points (Figure 5–2*a* and *b*). For either case, the scales of the *x* and *y* axes must be linear, which includes normal algebraic functions, trigonometric functions, and exponential functions, but not logarithmic functions.

For two known data points at (x_1, y_1) and (x_2, y_2), the *interpolant* (x, y) is obtained from:

$$y = y_1 + (y_2 - y_1)[(x - x_1)/(x_2 - x_1)]$$

---

[1] Definitions are from http://www.thefreedictionary.com/Houghton.

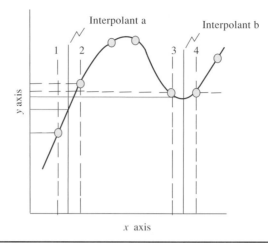

x axis

Figure 5-1 Schematic illustration of how meaningfully interpolating generates a value between known data points (shown by circles) depends on (1) the shape of the curve for the data and (2) how far the nearest known data points are from the interpolant. Here, the intervals between data points 1 and 2 and between data points 3 and 4 are the same, and an interpolant is to be found at the midpoint for each set. The effect of the shape of the curve that fits the data on the value of the interpolant found for each set is clear. The interpolant between points 3 and 4 is far less meaningful.

The accuracy of the new value for a true straight-line plot is very good. For cases where straight-line segments are used to approximate a nonlinear curve between known data points, interpolated values are less good, depending on how much the approximate straight-line segment differs from the actual curve.

Polynomial interpolation is a generalization of linear interpolation wherein the linear interpolant is replaced by a polynomial function of a higher degree than 1. Many times (especially today, using computers), regression analysis is used (or can be used) to generate such a curve-fitting equation. A curve-fitting polynomial of degree n has the form:

$$f(x) \text{ (or } y) = \pm ax^n \pm bx^{n-1} \pm cx^{n-2} \ldots \pm ix \pm j$$

As a rule, for a set of n data points, there is exactly one polynomial of degree $n - 1$ that goes through all the data points. The interpolant error is proportional to the distance between the data points to the power of n.

Figure 5–2c schematically illustrates a curve-fitting polynomial used for polynomial interpolation.

Spline interpolation uses low-degree polynomials in each interval between known data points, *and* the polynomial pieces are chosen such that they fit smoothly to abutting pieces. The resulting function is called a *spline*. For example, for the plot in Figure 5–2d, a natural cubic spline is piecewise cubic and twice continuously differentiable, with the second derivative being zero ($= 0$) at endpoints, generically being:

$$f(x) = \begin{cases} +a_1x^3 + b_1x^2 + c_1x + d_1, & \text{if } x \in [0,1] \\ +a_2x^3 + b_2x^2 + c_2x + d_2, & \text{if } x \in [1,2] \\ +a_3x^3 + b_3x^2 + c_3x + d_3, & \text{if } x \in [2,3] \\ \\ +a_4x^3 + b_4x^2 + c_4x + d_4, & \text{if } x \in [3,4] \\ +a_5x^3 + b_5x^2 + c_5x + d_5, & \text{if } x \in [4,5] \\ +a_6x^3 + b_6x^2 + c_6x + d_6, & \text{if } x \in [5,6] \end{cases}$$

As for polynomial interpolation, spline interpolation results in a smaller error than linear interpolation using straight-line segments.

An illustrative example should help.

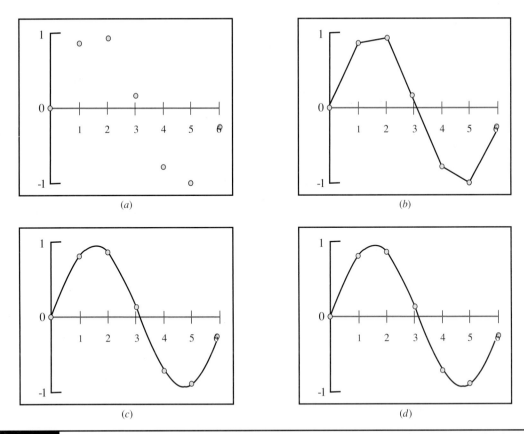

(a)

(b)

(c)

(d)

Figure 5-2 Examples of known data points simply plotted on a cartesian coordinate system, (in *a*) shown as a plot consisting of straight-line segments between pairs of known data points to approximate a curve to allow linear interpolation (in *b*); as a polynomial of order $n - 1$ (here, sixth-order) generated by regression analysis to pass through known data points (here, 7) to allow polynomial interpolation (in *c*); and as a curve created from low-degree or low-order (here, all cubic) polynomials to link pairs of known data points smoothly to allow spline interpolation (in *d*). (*Source:* Rendered after image from http://en-wikipedia.org/wiki/Interpolation; used freely under Creative Common Attribution 3.0 license, for which permission is not required.)

Illustrative Example 5-1: Using Interpolation Methods

Wikipedia provides a very nice example (at http://en.wikipedia.org/wiki/Interpolation, freely used under Creative Common Attribution free license) to illustrate linear, polynomial, and spline interpolation using the data points shown in Figure 5–2*a* and listed here as:

x	$f(x)$
0	0
1	0.8415
2	0.9093
3	0.1411
4	−0.7568
5	−0.9589
6	−0.2794

The three methods used to find an interpolant at $x = 2.5$ are as follows.

Using *linear interpolation,* and inserting values from the preceding list for $x_1 = 2$, $x_2 = 3$, and $x = 2.5$, with $y_1 = f(2) = 0.9093$ and $y_2 = f(3) = 0.1411$ into the operative equation:

$$y = y_1 + (y_2 - y_1)[(x - x_1)/(x_2 - x_1)]$$

gives for $y = f(2.5)$:

$$f(2.5) = 0.9093 + (0.1411 - 0.9093)[(2.5 - 2)/(3 - 2)] = 0.5252$$

Obviously, an easier approach in this case, where $x = 2.5$ is the midpoint between $x = 2$ and $x = 3$, would be to average the values of $f(2)$ and $f(3)$ to get:

$$(0.9093 + 0.1411)/2 = 0.5252$$

Using *polynomial interpolation,* for the given data set of seven points (i.e., $n = 7$), the computer-generated sixth-degree polynomial from regression analysis is:

$$f(x) = -0.0001521x^6 - 0.003130x^5 + 0.07321x^4 - 0.3572x^3 + 0.2255x^2 + 0.9038x$$

which, for $x = 2.5$ gives $f(2.5) = 0.5965$.

Using *spline interpolation* for the given data, here using all cubic polynomials that are each twice continuously differentiable with a second derivative of zero (0) at the ends, gives the following set of equations:

$$f(x) = \begin{cases} -0.1522x^3 + 0.9937x, & \text{if } x \in [0,1] \\ -0.01258x^3 - 0.4189x^2 + 1.4126x - 0.1396, & \text{if } x \in [1,2] \\ 0.1403x^3 - 1.3359x^2 + 3.2467x - 1.3696, & \text{if } x \in [2,3] \\ 0.1579x^3 - 1.4945x^2 + 3.7225x - 1.8381, & \text{if } x \in [3,4] \\ 0.05375x^3 - 0.2450x^2 - 1.2756x + 4.8259, & \text{if } x \in [4,5] \\ -0.1871x^3 + 3.3673x^2 - 19.3370x + 34.9282, & \text{if } x \in [5,6] \end{cases}$$

giving $f(2.5) = 0.5972$.

Both polynomial and spline interpolation tend to be computationally intensive compared to linear interpolation. Obviously, computers help to find the needed polynomials.

Extrapolating

The technique of extrapolating new points outside the range for which data points or values are known (from calculation, measurement, or sampling) results in values that are subject to greater uncertainty than typically is associated with interpolating between data points within the range. Many times (if not most times), extrapolation is accomplished graphically using actual data plots (see "Graphing" in Chapter 26) and not by calculation or estimation using numbers from these plots. This is not the case with interpolation.

Extrapolation methods include:

■ Linear extrapolation
■ Polynomial extrapolation
■ Conic extrapolation
■ French-curve extrapolation

The choice of one method over another relies on a priori knowledge of the process by which the existing data points were created. Figure 5–3 schematically illustrates how a meaningful value is sought for the box, given data points shown by round dots.

Linear extrapolation is most accurate when used to extend the graph of an approximately linear function or to find a new value near the last known data point(s) at the end of the known data range for a nonlinear

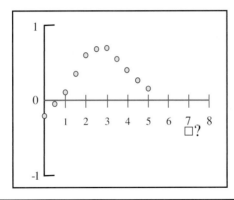

Figure 5-3 An example of the problem of extrapolating a meaningful value for the box using data given by the round dots. (*Source:* Rendered after image from http://en-wikipedia.org/wiki/Interpolation; freely used under Creative Common Attribution 3.0 license, for which permission is not required.)

curve. The physical process, when performed graphically, involves creating and extending a tangent line at and from the end of the known data. For a plot consisting of n data points, a new data point at x^* can be extrapolated from the two nearest known data points at (x_{n-1}, y_{n-1}) and (x_{n-2}, y_{n-2}) by the equation:

$$y(x^*) = y_{n-1} + [(x^* - x_{n-1})/(x_n - x_{n-1})] \, (y_n - y_{n-1})$$

This equation is identical to that for linear interpolation if $x_{n-1} < x^* < x_n$.

In the process of linear interpolation, it is possible to include more than two known data points and average the slope of the linear interpolant.

Polynomial extrapolation involves extending a polynomial curve beyond known data and requires more advanced mathematics (e.g., Lagrange interpolation or Newton's method of finite differences to create a Newton series that fits the data) than is covered by this book. The interested reader is encouraged to seek a specific reference source, perhaps on the Internet under "polynomial extrapolation."

Conical extrapolation involves creating a *conic section* (i.e., circle, ellipse, parabola, or hyperbola) using five data points near the end of the known data range to extend the curve. Using conic sections mathematically requires recognizing the form of the equations (Chapter 3) for circles, ellipses, parabolas, and hyperbolas. Conical extrapolation can be done using a conic section template on paper (as a graphical technique; see discussion of templates in Chapter 25). It can also be done with a computer.

French-curve extrapolation generally uses a French-curve template for data distributions that tend to be exponential. The actual process can use an actual French-curve template, as used in mechanical drawing or drafting, or a computer.

Again, an illustrative example for extrapolation should help.

Illustrative Example 5-2: Using Linear Extrapolation

Once again, as for Illustrative Example 5-1, Wikipedia has a great example illustration for linear extrapolation (used under Creative Common Attribution free license).

Using the data in Figure 5-3, the problem is to assign a meaningful value at the box at x = 7, given the data points shown by dots.

The two nearest data points on the plot are at x_n = 5 and x_{n-1} = 4.5, for which the estimated values (by scaling) on the y axis are y_n = 0.154 and y_{n-1} = 0.269, respectively.

Inserting these values into the operative equation

$$y(x^*) = y_{n-1} + [(x^* - x_{n-1})/(x_n - x_{n-1})]\,(y_n - y_{n-1})$$

and solving for $y(x^*)$, where $x^* = 7$, yields:

$$y(7) = 0.269 + [(7 - 4.5)/(5 - 4.50]\,(0.154 - 0.269) = -0.306$$

A good example of where interpolation and/or extrapolation would be needed is to predict the horizontal distance of flight (and impact point) for a projectile, if one has data for various firing angles.

In Summary

Interpolating and *extrapolating* allow values of a dependent parameter (or output) to be created either between or beyond values for known independent parameters (in inputs), respectively. The accuracy of the new values depends on the linearity versus nonlinearity of the known data plot and how far the new point is to be from nearest known data points. Graphical approaches are more common for extrapolation, but mathematical (analytical) approaches are available for both interpolating and extrapolating.

Suggested Resources

http://www.en.wikipedia.org/wiki/Interpolation
http://www.en.wikipedia.org/wiki/Extrapolation

CHAPTER 6

Checking

ENGINEERING DEMANDS CARE and accuracy. The consequences of carelessness and/or inaccuracy range from annoyance at a part that doesn't fit properly into an assembly to a catastrophic disaster that can be seen by the world as a sad testimony to sloppy engineering. An example is the collapse of the Tacoma Narrows Bridge.

Spanning the Tacoma Narrows, a strait of Puget Sound between the city of Tacoma south of Seattle and the Kitsap Peninsula in the state of Washington, the Tacoma Narrows Bridge was the third-longest-span suspension bridge (behind the Golden Gate Bridge in San Francisco and the George Washington Bridge in New York City) when it was opened on July 1, 1940. From the time the deck was built, it began to move vertically in windy conditions, which were common in the Narrows. This behavior led construction workers to give the bridge the nickname "Gallopin' Gertie."

The bridge collapsed under a 40 mph (64 km/h) wind on November 7, 1940, just four months after it opened. The cause of the collapse was elementary *forced resonance* brought on by aerodynamic flutter; things good engineers knew about from their undergraduate educations.

Because the bridge swayed violently for some time before it collapsed, no one was killed. However, the disaster became an albatross around the neck for engineers because of the shameful notoriety it drew.

A remarkable 16mm movie taken on the day of the disaster can be seen on YouTube.com under a search for "Tacoma Narrows Bridge Collapse Gallopin' Gertie." Still images are all copyrighted, as they are irreplaceable, and permissions are not granted for free!

The message is: Engineers need to be careful, thorough, and accurate when they solve problems—even easy problems. An important technique to use is to check one's work at all stages during design and at each step of problem-solving.

Checking, as it pertains to problem-solving, according to the dictionary, is defined as[1] "the action or an instance of inspecting or testing, as for accuracy or quality." Checking can and should be used for all kinds of work by engineers, not just problem-solving, and, within problem-solving, not just for mathematical approaches.[2]

When an engineer has the advantage (increasingly, the luxury) of being able to have another (often more experienced) engineer check his or her work, this is great. However, it is only common sense that one should *never* let checking be done *only* or *first* by such a person, particularly not a supervisor or a client. It is unwise—and unprofessional—to submit work to a supervisor or client that one has not first checked himself or herself during the process of accomplishing the work.

Checking is particularly important—and easy to do—when mathematics is involved. The reason is simple: Mathematics almost always leads to a single, quantitative solution to an equation or set of equations, that is, to a closed solution. The general approach is to use the calculated value as one of the known parameters and one of the originally known parameters as the unknown, and repeat calculation for the original equation rearranged to solve for the new "unknown." If the initial calculation of the

---

[1]Houghton-Mifflin's online dictionary at http://www.free dictionary.com/Houghton.
[2]There is an adage used by carpenters that should equally well be heeded by engineers: "Measure twice. Cut once." In the case of engineers, besides measuring twice, they should calculate twice!

original unknown was done correctly, the value obtained for the given parameter acting as a new "unknown" will match the originally given value.

A couple of illustrative examples should make clear how (and why) checking can (and should) be done by an engineer on his or her own.

Illustrative Example 6-1: Checking a Simple Equation

The kinetic energy KE of an object of mass m moving at a constant velocity v is given by

$$KE = 1/2\ mv^2$$

If values for mass and velocity of an automobile are given as m = 1430 kg and 80.0 km/h (kph), the kinetic energy in joules, where 1 J is 1 kg-m^2/s^2), is:

$$KE = 1/2\ (1430\ kg)[(80.0\ km/h)(1000\ m/km)(1\ h/3600\ s)]^2$$
$$= 353,086.4198\ J\text{—on a calculator}$$

which, to three significant figures (from the value of the velocity) is:

$$KE = 353,000\ J,\ or\ 353\ kJ$$

To check this (or any calculation!), work the problem another way, solving for one of the given parameters—say, velocity v, assuming the mass m and the kinetic energy KE are known, using a rearranged version of the operative equation, here:

$$v = (2\ KE/m)^{1/2}$$

Thus,

$$v = \{(2)[(353,000\ kg\text{-}m^2/s^2)/1430\ kg]\}^{1/2} = 22.2\ m/s$$

which, converted to km/h, is:

$$v = (22.2\ m/s)(3600\ s/h)(1\ km/1000\ m) = 79.9\ kph$$

This is certainly close enough to the known value of 80 kph to ensure that calculation of KE was done correctly. In fact, to two significant figures, as set by the given value of velocity v, the answer—80 kph—is identical to the given value. Being so simple to do, there is no reason for an engineer (no less a student) not to check his or her math.

Illustrative Example 6-2: Checking a More Complicated Equation

It is easy to make a mistake when working with exponential equations. The two places mistakes arise most often are: (1) in dealing with + or − exponential factors or arguments[3] between denominators and numerators and (2) in rearranging the equation.

The equation for the rate at which a given atomic species (i.e., solute atoms) moves by diffusion D (which involves atomic-level jumps) in a solid crystalline host (i.e., a solvent) is given by the exponential (Arrhenius) equation:

$$D = D_o\ exp\ (-Q_d/RT)$$

in which D_o is a preexponential, temperature-independent constant with the same units as D (i.e., m^2/s), Q_d is the activation energy (i.e., energy barrier) for the diffusion of 1 mole of solute to take place (in units of J/mol), T is the temperature in absolute degrees (K), and R is the gas constant (= 8.31 J/mol-K).

---

[3]The argument is what appears within a logarithm or an exponent.

To calculate a value for Q_d if D, D_o, and T are known, the properly rearranged equation is:

$$Q_d = -RT \ln (D/D_o)$$

A careful engineer (or engineering student) would first check the units of the rearranged equation to see if they yield the proper units for Q_d, here:

$$(J/mol\text{-}K)(K) \ln [(m^2/s(/(m^2/s)] = J/mol$$

The correct units for Q_d indicate that the equation has been properly rearranged. As is always the case, the units in the argument of a logarithm or an exponential must cancel to yield a dimensionless number.

Inserting values of $D = 5.8 \times 10^{-13}$ m^2/s, $D_o = 1.2 \times 10^{-4}$ m^2/s, and $T = 550°C$ (823 K), one gets Q_d as:

$$Q_d = -(8.31 \text{ J/mol-K})(823 \text{ K}) \ln [(5.8 \times 10^{-13} \text{ m}^2/s)/(1.2 \times 10^{-4} \text{ m}^2/s)]$$
$$Q_d = -(8.31 \text{ J/mol-K})(823 \text{ K}) \ln [4.833 \times 10^{-9}]$$
$$Q_d = -(6839 \text{ J/mol})(-19.15) = 130{,}967 \text{ J/mol} = 131 \text{ kJ/mol}$$

A common mistake for students, in particular, is to erroneously calculate $5.8 \times 10^{-13}/1.2 \times 10^{-4}$ to be 4.833×10^{-17} (really, 4.8×10^{-17}, considering significant figures), by not properly switching the sign of the power of 10 in the argument of the exponential in the denominator when moved into the argument of the exponential in the numerator.

This mistake would give Q_d as:

$$Q_d = -(6{,}839 \text{ J/mol}) \ln (4.833 \times 10^{-17}) = -(6839 \text{ J/mol})(-37.56)$$
$$Q_d = 257 \text{ kJ/mol}$$

Checking by calculating the known value of D using the calculated value of Q_d would allow the error to be found, thus:

$$D = D_o \exp (-Q_d/RT)$$
$$D = 1.2 \times 10^{-4} \text{ m}^2/s \exp (-257{,}000 \text{ J/mol})/(8.31 \text{ J/mol-K})(823 \text{ K})$$
$$D = 1.2 \times 10^{-4} \text{ m}^2/s \exp (-37.58) = 4.88 \times 10^{-17} \text{ m}^2/s$$

which is wrong, as the given value was 5.8×10^{-13} m^2/s.

In fact, it is off (at 10^{-17} versus 10^{-13}) by four orders of magnitude, which suggests a four-order-of-magnitude error in the argument of the exponential.

In Summary

Beyond being professional, it is common sense for engineers to check their work. *Checking* should be done for each step within each stage of a job (so as not to carry forward an error that, by compounding, will be harder to find later) and for all types of problems, not just those involving mathematical equations. The process is simple for math—that is, work the equation to solve for one of the given (known) parameters by inserting the just-calculated value of the original unknown into the rearranged equation. When the calculated value is the same as the originally given parameter, the math is correct. However, checking can and should be done for all techniques and/or processes by which a problem is solved, not just for mathematical approaches.

CHAPTER 7

Dimensional Analysis and Reconciling Units

IN CHAPTER 3, "Using Equations," it was pointed out that one of the essential factors to be considered before using mathematical equations to solve an engineering problem is to analyze and reconcile the units. In this chapter, the importance of dimensional analysis and reconciling units as a problem-solving technique is considered on its own account.

In the context of this book, at least, the difference between dimensional analysis and reconciling units is that:

- *Dimensional analysis* involves looking at the units of the input parameters in an equation and of the output parameter of (or solution to) the equation to get a sense of the physical meaning of each and of the equation from its form. (This is used in Chapter 11, "Sensitivity Analysis.")
- *Reconciling units* involves making input parameters self-consistent so that all will be in the same system (e.g., metric kgs or cgs, or English) so as to generate the needed or desired units for the output of or solution to the equation.

These two subtly different techniques should always be used together, as they are both parts of the same goal; to ensure that solution of the equation will result in the needed or desired outcome.

The techniques of *dimensional analysis* and *reconciling units* should be considered for the following relating to problem-solving:

1. Dimensional analysis should be used to check that an equation has been written correctly, both in the reference source, where errors can be made by authors or publishers, or in transcription by the user. This is particularly important to do after an equation has been rearranged from some standard or original form (e.g., $E = mc^2$ used to solve for m from E and c, the speed of light in a vacuum, as $m = E/c^2$). (In this case, dimensional analysis is a *checking technique,* discussed in Chapter 6.)
2. Reconciling units should be used to yield the needed or desired units for the solution to an equation from the input parameters, often requiring the use of conversion factors. (This could also be part of checking.)

and, while not often recognized or used enough,

3. Dimensional analysis can be used to reconstruct a vaguely recollected equation, working upward from first principles of physics.

The remainder of this chapter presents illustrative examples for each of these possible uses of dimensional analysis and reconciling units.

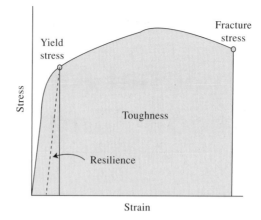

Figure 7-1 The area under a stress-strain curve is the energy a material can absorb. The mechanical property is known as toughness when the integral is taken to the strain under loading just before fracture and is known as resilience when the integral is taken to the strain under loading just before yielding.

Illustrative Example 7-1: Dimensional Analysis for Physical Significance

For engineers, in general, and for engineering students, in particular, mathematical equations must be understood (as opposed to memorized!) for their physical significance.[1] Analyzing the dimensions in the equation often helps to understand the significance of the various input parameters, their arrangement—and, thus, interplay—in the equation (i.e., the form of the equation), and the solution to or output of the equation.

An important mechanical property for a material used to make a structural element that is to be subjected to impact (e.g., the landing gear of an airplane) is *toughness*. The toughness of a material is its ability to absorb energy without failure. It is represented by the area under a stress-strain (or strength versus elongation) diagram or plot (Figure 7-1).[2]

The area under the stress-strain curve is the product of these two parameters (often by integration), so the units are:

$$\text{Stress (MPa = N/m}^2) \times \text{strain (m/m)} = \text{toughness (N-m/m}^3)$$

or, in English units:

$$\text{Stress (lbm/in}^2) \times \text{strain (in/in)} = \text{toughness (in-lbm/in}^3)$$

This analysis of dimensions makes it clear that toughness is the energy per unit volume the material can safely absorb.[3]

---

[1] If there is a general approach that describes what engineers do, it is that they represent physical systems in terms of mathematical equations to allow manipulation of (or operation on) the equation to optimize the physical system by optimizing the mathematical solution.

[2] The energy that a material can absorb is given by the integral of stress versus strain, $\int \sigma \, d\varepsilon$. Taken from 0 to ε_f just before fracture (under loading), the property is known as *toughness*. Taken from 0 to ε_y just before yielding (under loading), the property is known as *resilience*. The latter energy is all recoverable, having been stored elastically, as in a vaulting pole.

[3] Notice how, during dimensional analysis, it would be misleading to reduce the dimensions of the product of stress and strain, e.g., N-m/m^3 ⇒ N/m^2, the latter units being those for stress alone. The key is that strain is not dimensionless. Strain is the change in length per unit length of a structural element subjected to tension or compression, e.g., mm/mm, m/m, in/in. These are all equivalent, but that doesn't make them dimensionless for use in dimentional analysis (e.g., mm/mm is not dimentionaless).

Illustrative Example 7-2: Dimensional Analysis to Check a Rearranged Equation

The electrical conductivity σ_{el} of a material may be expressed as

$$\sigma_{el} = n \mid e \mid \mu_e$$

in which n is the number of conducting electrons available per unit volume of the material (m^{-3}), $\mid e \mid$ is the absolute magnitude of the charge on an electron, a physical constant (= 1.6×10^{-19} C), and μ_e is the mobility of an electron in the material (in units of m^2/V-s). With a coulomb (C) having units of amperes/second (C = A/s), an ohm of resistance having units of V/A (Ω = V/A), and a reciprocal ohm being Ω^{-1} = A/V, if this equation (in standard form) is rearranged to allow calculation of μ_e, dimensional analysis will show whether the rearrangement was done correctly, thus:

$$\mu_e = \sigma_{el}/n \mid e \mid$$
$$\mu_e \text{ (to be } m^2/\text{V-s)} = (\Omega^{-1}\text{-}m^{-1})/(m^{-3})(C) \, Þ \, (A/V)(m^{-1})/(m^{-3})(A\text{-}s)$$
$$(A/V)(m^{-1})/(m^{-3})(A\text{-}s) \, Þ \, m^2/\text{V-s}$$

So the rearranged equation is correct!

Illustrative Example 7-3: Reconciling Units to Yield Desired Units

When a structural material (or structural member made from some material) contains a flaw (e.g., a surface or internal crack), a critical stress σ_c will be reached at which a flaw of given size (of length a for a surface or edge flaw, $2a$ for an internal flaw) will suddenly propagate to cause complete fracture of the material or structural member. The magnitude of the critical stress is dependent upon the inherent fracture toughness of the material K_{IC}. The operative equation for σ_c is:

$$\sigma_c = K_{IC}/Y(\pi a)1/2$$

in which Y is a geometric factor to account for the width-to-thickness ratio of the structural member, with a typical value of 1.0 and no units.

Figure 7-2 shows a structural member containing an internal flaw of length $2a$ (at the left) and a surface or edge flaw of length a (at the right).

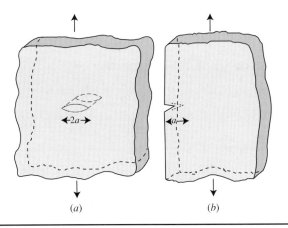

(a)   (b)

Figure 7-2 Schematic representation of (a) an internal flaw in a thick plate of material with infinite width and (b) for a surface or edge flaw in a thick plate of material with semi-infinite width. Such widths are necessary to yield the material property of K_{IC} under conditions of plane strain. The factor Y in the equation for K_{IC} accounts for widths different than these.

Assume you must use this equation to calculate σ_c in units of ksi (i.e., 10^3 lbm/in^2) for a surface flaw, but have the other parameters involved in the equation in mixed units, as follows: $K_{IC} = 60$ MPa(m$^{-1/2}$), $a = 0.020$ in, $Y=1$, and $\pi=3.14$. Obviously, you have to reconcile units using appropriate conversion factors (see Chapter 3) to generate the desired units for σ_c as follows:

$$\sigma_c \text{ lbm/in}^2) = (60 \text{ MPa-m}^{1/2})(1 \text{ ksi-in}^{1/2}/1.099 \text{ MPa-m}^{1/2})/(1.0)(3.14)^{1/2}(0.020 \text{ in})^{1/2}$$

yielding, after cancellation of like units:

$$\sigma_c = 217.9 \text{ ksi}$$

or about 220 ksi to two significant figures.[4]

Illustrative Example 7-4: Dimensional Analysis to Reconstruct an Equation

You will find, after being out of engineering school for many years, and/or having not used a particular equation for a long time, that you may not remember an equation you require to solve a problem. If you have access to a computer and the Internet, you'll find it. If not, it will help greatly if you remember the origin of the equation from first principles. If you learn properly, you will learn the first principles (i.e. the underlying physics) rather than just a mathematical equation.

Presume you need to calculate the density of a metal ρ knowing its atomic weight A and, from some work done by you using x-ray diffraction, for example, its lattice parameter a_o. You vaguely recall that density represents how tightly packed atoms of a material are per unit volume. You know the weight (or mass) of the atom from its atomic weight, being molybdenum Mo, as $A = 95.94$ amu per atom or g/mol. You know from your x-ray diffraction work that the lattice parameter $a_o = 0.306$ nm for Mo, which is the length of the side of the cubic unit cell that, by translation along its three orthogonal axes, allows space to be filled with Mo atoms. You either recall or learned from your x-ray diffraction work that Mo has a body-centered cubic (BCC) crystal structure for which there are two atoms of Mo per unit cell—in other words, eight atoms at corners of a cube, of which 1/8 of each lies within—so belongs to—the cube, plus one atom at the center of the cube, or $8 \times 1/8 + 1 = 2$ (Figure 7-3). You recognize that you need to arrange the number of atoms per unit cell n, the volume of the unit cell V_c (in nm^{-3}), the atomic weight A (in g/mol), and, to deal with the moles of Mo, Avogadro's number $N_A = 6.023 \times 10^{23}$ atoms/mol in such a way that you get units of density ρ in g/cm^3 (grams per volume). Simplifying, n has units of atoms per unit cell, V_c has units of volume, A has units of grams per mole, and N_A has units of atoms/mole.

With a little playing around, guided by the units alone for each input parameter, you end up with:

$$\rho = nA/V_cN_A \Rightarrow$$
$$\text{(atoms/unit cell)(g/mol)/(nm}^3\text{)(cm}^3\text{/nm}^3\text{)(atoms/mol)} \Rightarrow \text{g/cm}^3$$

Inserting the known values, knowing that $V_c = a_o^3$, gives:

$$\rho = (2)(95.94 \text{ g/mol})/(0.312 \text{ nm})^3(1 \text{ cm}/10^7 \text{ nm})^3(6.023 \times 10^{23} \text{ mol}^{-1})$$
$$\rho = 10.4 \text{ g/cm}^3$$

This compares well with literature values of 10.22 g/cm^3.

[4]This value is just about the yield strength of the quenched and tempered AISI 4140 steel for which the value of K_{IC} was taken. This indicates that a structure made from this material could tolerate a surface flaw 0.020 in long so long as the applied stress never reached the yield strength of 218 ksi, which safe design practice would ensure never occurred by employing a factor of safety on the allowable maximum value of stress.

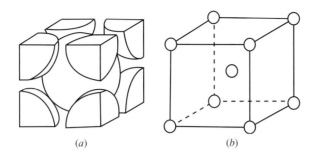

(a)   (b)

Figure 7-3 Schematic representation of what is known as a body-centered cubic (BCC) crystal structure, for which the repeat unit cell has a Mo atom at each corner and at the center. However, the actual number of Mo atoms that belong entirely to the unit cell is two, as shown at the left of the figure by the sharing of corner atoms by eight surrounding cells and the central atom that belongs entirely to the unit cell.

In Summary

The techniques of *dimensional analysis* and *reconciling units* are to be used together in solving engineering problems using mathematical equations. They allow the equation to be checked for correct transcription from some reference source (or memory), and with understanding the physical significance of the various input parameters, the form of the equation itself, and the solution to the equation, allow a rearranged equation to be checked for proper rearrangement, and even allow vague equations to be reconstructed using units for the input parameters so that they yield the correct units for the solution.

CHAPTER 8

Using Similarity and Ratios

EVERY ENGINEER AND every engineering student should be familiar with the mathematical concept of similarity from plane geometry in secondary school. In its simplest form, *similarity* means that two things have the same shape. All circles are similar to each other, just as all squares are similar to each other and all equilateral triangles are similar to each other. In fact, each of these 2D geometric shapes, besides being similar within a type, is congruent within that type. Two shapes are *congruent* when one is created from (or related to) another by a uniform scaling or proportionality factor, enlarging or shrinking the one to produce the other. In this way, polygons are both similar and congruent if corresponding sides are in proportion and corresponding angles have the same measure. One can be obtained from the other by uniformly stretching (or shrinking) by the same amount in all directions. In this way, two teapots could be similar and congruent, whether in 2D silhouette (Figure 8–1) or 3D, even though one has the same shape as the mirror image of the other.

Figure 8–2 shows similar shapes by like cross-hatching or shading. Note from these examples that ellipses are *not* all similar (if the ratios of major to minor axes are different), nor are rectangles all similar (if aspect ratios for long-to-short sides are different).

Triangles are an especially interesting case, you hopefully will recall, as three criteria are sufficient to prove two triangles are similar: (1) angle-angle (AA) or angle-angle-angle (AAA), (2) side-side-side (SSS or SSS~), or (3) side-angle-side (SAS or SAS~). Remember? In summary:

(a)

(b)

(c)

Figure 8-1   The teapots shown in these silhouettes are all geometrically similar and congruent, since one can be produced from another by shrinking or stretching (*a* and *b*), even if rotation or reflection is involved (*c*).

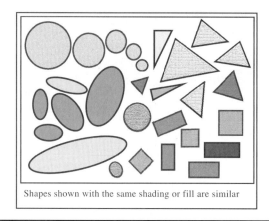

Shapes shown with the same shading or fill are similar

Figure 8-2 Like-shaded or cross-hatched shapes are geometrically similar.

- Two triangles are similar if they have two corresponding (i.e., pairs) or angles with the same measure, as specifying two angles fixes the third angle since the sum of all three angles in a triangle is 180°.
- Two triangles are similar if the ratio of all three corresponding sides for the two is the same.
- Two triangles are similar if two corresponding sides have the same ratio and if the angle between these sides (i.e., the included angle) has the same measure.

Triangles are particularly important in structural engineering because, as a frame, for example, they are stable against shearing forces and shape change. This is taken advantage of in structural engineering in the form of trusses (Figure 8–3).

In its 3D form, we are all familiar with geometric similarity being used by ancient engineers, as in the pyramids of Egypt (Figure 8–4), whether to scale engineering calculations or for aesthetic purposes.

(a)   (b)

Figure 8-3 Triangles are used in structural trusses since they offer shape stability against shear. Prefabricated wood trusses for residential home and small building framing (*a*) and steel girder trusses in a railroad bridge (*b*) are shown. (*Sources:* [a] from K-Wood Truss Rafters, Inc., Grantsburg, Wisconsin; used with permission; [b] from Wikimedia, originally submitted by Leonard G. on 23 January 2006, under Creative Commons ShareAlike 1.0, used under free license.)

Figure 8-4 One can see geometric similarity in the pyramids at Giza, Egypt. While the largest, the Great Pyramid of Khufu (centermost of the three larger pyramids in the background), is best known, its two slightly smaller neighbors are also familiar from many photographs. The much smaller pyramids in the foreground (two of which are stepped types) are similar in key dimensions, such as base perimeter-to-altitude ratio and slope angle. (*Source:* Wikimedia Commons; originally contributed by Riclib on 17 June 2007, used freely under Creative Commons Attribution-Share Alike 2.0 Generic license.)

The case of similarity between the teapots shown in Figure 8–1 derives from similarity in euclidean space, being termed *similarity transformation* or *dilation*, in which all distances in space are multiplied by the same positive scalar *r*. Mathematically, for any two points *x* and *y*, this is expressed as:

$$d(f(x), f(y)) = r\, d(x,y)$$

in which $d(x,y)$ is the euclidean distance from *x* to *y*. This too can be extended to 3D space.

What has been discussed so far is known as *geometric similarity*. Geometric similarity becomes especially valuable as an engineering problem-solving technique in two contexts:

1. In *variant design*, in which a new design is motivated by and, often, largely derived from another earlier or preexisting design by scaling up or down from the original design. (This could be as straightforward as doubling the capacity of a thoroughfare by doubling the number of lanes, but it could also apply to increasing the passenger-carrying capacity and/or range of a commercial airplane by scaling the fuselage, wings, horizontal and vertical stabilizers, etc.)[1]
2. In moving between *models* and *prototypes* (see Chapter 31) in which true models at least reproduce key features of the prototype but on a scale that is usually (but not always) smaller. Geometric similarity exists between a model and a prototype if the ratio of all corresponding dimensions in the model and the prototype are equal.

    For length:

$$L_{\text{model}}/L_{\text{prototype}} = L_{\text{m}}/L_{\text{p}} = \lambda_{\text{L}}$$

where λ_{L} is the scale factor for length.

---

[1]One can imagine that some variant designs involve more than geometric similarity, and require new materials, new processes, and/or new technologies in order to increase or decrease the scale of a similar design. For example, laptop computers might be seen as little brothers or sisters of desktop PCs, but they may well demand more than geometric scaling. Likewise for tiny microphones, tiny in-the-ear hearing aids, or much taller skyscrapers.

For area:

$$A_{model}/A_{prototype} = L_m^2/L_p^2 = \lambda_L^2$$

All corresponding angles are the same.)

Illustrative Example 8-1: Using Geometric Scaling

The small company you work for as the only engineer is growing in business and, therefore, needs more space. The owner has decided to contract for "a geometrically similar building" to the one the company presently occupies. The current one-story building has a floor plan of 80 ft × 160 ft with a 10-ft ceiling. The new building is to be "25 percent bigger in all dimensions." Out of curiosity, you decide to look at how much more floor area and volume a "25 percent bigger" building will have, as floor area is usable space and is the basis for estimating the cost of many standard commercial buildings (see Chapter 4), while the increase in volume due to a higher ceiling will add to heating, ventilation, and air-conditioning (HVAC) costs without adding to usable space.
Your calculations for geometric similarity show:

Present Building	**New Building**
Floor area: (80 ft)(160 ft) = 12,800 ft²	[(1.25)(80 ft)(1.25)(160 ft)] = 20,000 ft²

or 56.25 percent greater usable floor area; $(1.25)^2 = 1.5625$.

Volume: (80 ft)(160 ft)(10 ft) = [(1.25)(80 ft)(1.25)(160 ft)
128,000 ft³   (1.25)(10 ft)] = 250,000 ft³

or 95.3 percent greater volume and higher HVAC costs; $(1.25)^3 = 1.953$.
Scaling up the floor plan seems fine, but the ceiling should remain at 10 ft. Doing this, the present and new buildings will *not* be geometrically similar—but HVAC costs won't be unnecessarily high.

Similarity in engineering and engineering problem-solving goes beyond geometric similarity, however. Without attempting to be comprehensive, two important examples are: (1) *kinematic similarity* and (2) *dynamic similarity*.

Kinematic similarity involves (or requires) similarity of time *t* as well as of geometry *L*. It exists between a model (m) and a prototype (p), for example, when:

1. The paths of moving particles are geometrically similar.
2. The ratios of the velocities *v* and accelerations *a* of particles are similar, so that:

$$v_m/v_p = (L_m/t_m)/(L_p/t_p) = \lambda_L/\lambda_t = \lambda_v$$
$$a_m/a_p = (L_m/t_m^2)/(L_p/t_p^2) = \lambda_L/\lambda_t^2 = \lambda_a$$

Kinematic similarity is useful in that streamline patterns (for aerodynamics or hydrodynamics) are the same.

Illustrative Example 8-2: Using Kinematic Similarity

While sitting in your car as a long freight train rolls past a railroad crossing in your town, your mind wanders. You notice that the boxcars are labeled as 50 ft long, and it takes just over a second for one to pass a fixed point. Remembering from a favorite science teacher in eighth grade that 60 mph is 88 ft/s, you estimate that the train is traveling at about 30 mph (i.e., about 44 ft/s). Your mind flashes back to the HO-gauge freight train you had as a preteen. You wonder how fast you ran your model train compared to how fast this real train is traveling.

Your track layout was a simple oval with 2-ft-radius ends and 5-ft-long straight sections, and you remember once timing your HO train at 14 s for one complete loop of the oval.

The scale for HO trains is 1:87, which represents the geometric similarity between the model train and a real train (or model and prototype). The length of one loop of your oval layout was:

(2 semicircular ends)($\pi d/2$ ft per end) + (2)(straight section ft) =
(2)[(3.14)(4 ft)/2] + (2)(5 ft) = 22.6 ft

This would mean that one loop around your oval track layout would be equivalent to 87 × 22.6 ft = 1970 ft for a real train.

At the speed you ran your model train, a real train would have to travel 1970 ft in 14 s. Converted to miles per hour, this would be:

(1970 ft/5280 ft/mi)(14 s)(1 h/3600 s) = 96 mph!

For your HO train to be kinematically similar to the real train you are watching, the time for one loop of your model layout would have to have been

(96 mph/30 mph)(14 s) = 45 s

That might have been realistic, but it would have been boring!

Dynamic similarity involves (or requires) geometrically and kinematically similar systems, provided the ratios of all forces F on the model and on the prototype are the same. Generally,

$$F_m/F_p = m_m a_m/m_p a_p = \rho_m L_m^3/\rho_p L_p^3 \times \lambda_L/\lambda_t^2 = \lambda \rho \lambda_L^2 (\lambda_L/\lambda_t)^2 = \lambda \rho \lambda_L^2 \lambda_v^2$$

This occurs when the controlling dimensionless group on the right side of the previous defining equation is the same for the model and the prototype. In aerodynamics and hydrodynamics, such a dimensionless parameter is the *Reynold's number* (see Chapter 14).

Similarity, in all contexts described, involves ratios or proportions or scaling. But, as a problem-solving technique in engineering, there are other situations in which using ratios is valuable. First, there is the so-called golden ratio found in nature and used in engineering and art. Second, there are mathematical equations that describe physical systems and/or phenomena that, at first glance, may not appear sufficient to allow a solution to be calculated. It may appear, at first glance, that there are too many unknowns. A little reflection, however, along with a return to some basic mathematical principles, often reveals that the route to success is to use ratios.

The golden ratio is a particularly interesting—in fact, fascinating—ratio that has found its way into geometry, art, architecture, and other areas, including engineering. The *golden ratio*, usually symbolized by the lowercase Greek letter phi (ϕ) has a value of 1.618, but as an irrational mathematical constant can be written to thousands of decimal places, typically, though, written as 1.6180339887 . . . . The golden ratio (or golden mean) was first recognized by ancient Greek mathematicians more than 2400 years ago, with its frequent appearance in geometry.[2]

The simplest idea behind the golden ratio is that it appears if you divide a straight line into two parts so that the length of the longer part divided by the length of the shorter part is equal to the whole length divided by the longer length (Figure 8–5).

[2]Greek mathematicians were especially intrigued by the golden ratio for its appearance with regular pentagrams (that look like the five-pointed star we all drew as children) and regular pentagons, with either one being able to be inscribed within the other.

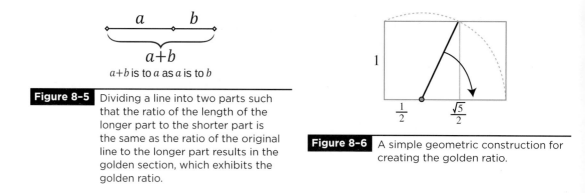

Figure 8-5 Dividing a line into two parts such that the ratio of the length of the longer part to the shorter part is the same as the ratio of the original line to the longer part results in the golden section, which exhibits the golden ratio.

Figure 8-6 A simple geometric construction for creating the golden ratio.

One can also calculate the value of the golden ratio by starting with any number x and following these steps:

1. Divide 1 by the number x (i.e., $1/x$).
2. Add 1 to the result.
3. Making this the new number x', start over at Step 1 and repeat all steps over and over.

With an electronic calculator, just keep pressing "$1/x$," "+," "1," and "=" over and over, starting with $x = 2$ to get:

Number	1/Number	Add 1
2	1/2 = 0.5	0.5 + 1 = 1.5
1.5	1/1.5 = 0.666 . . .	0.666 . . . + 1 = 1.666 . . .
1.666 . . .	1/1.666 . . . = 0.6	0.6 + 1 = 1.6
1.6	1/1.6 = 0.625	0.625 + 1 = 1.625
1.625	1/1.625 = 0.6154	0.6154 + 1 = 1.6154
1.6154	etc.	

Keep going for a long time and the result will be the golden ratio . . . to thousands of decimal places, if one is patient and persistent enough!

The value of ϕ can also be generated with simple geometry, thus (Figure 8–6):

- Draw a unit square (with sides = 1).
- Place a dot at the midpoint of one side.
- Draw a line from this dot to the opposite corner (so the length of the new line will be $\sqrt{5}/2$).
- Rotate the new line downward until it reaches the line with the dot at its midpoint.

You end up extending the original unit square into a rectangle with the golden ratio between its long and short sides, for which

$$\phi = 1/2 + \sqrt{5}/2 = (1 + \sqrt{5})/2 \sim 1.618034$$

Throughout the ages, artists and architects have thought that the golden ratio creates the most pleasing and beautiful shapes (Figure 8–7). But there may be more to it!

Most fascinating of all is this surprise. If one takes two successive Fibonacci numbers,[3] their ratio

[3]Fibonacci numbers result when any two numbers are added, and then the new sum is added to the preceding number, etc. The series often is started with 0 and 1 to produce 1, then 1 + 1 to produce 2, then 1 + 2 to produce 3, then 2 + 3 to produce 5, then 3 + 5 to produce 8, etc. Fibonacci numbers, however, can start with any pair of numbers—say, 18 and 243 to give 261, which, with 243, gives 504, etc.

Figure 8-7 The golden ratio appears throughout Greek architecture and art, as is shown here, with a superimposed golden rectangle, for the Parthenon, built from 447 to 438 BC. The temple was dedicated to the Greek goddess Athena, patron goddess of Athens. (*Source:* WikiMedia, originally contributed by Dimboukas on 21 October 2011, and used under Creative Commons free license without requiring permission.)

is close to the golden ratio, and, in fact, their ratio gets closer and closer to the golden ratio as the Fibonacci numbers get larger and larger; thus:

A	B	B/A
2	3	1.5
3	5	1.666 . . .
5	8	1.6
8	13	1.625
–	–	–
144	233	1.618055556 . . .
233	377	1.618025751 . . .

It turns out that Mother Nature creates many of her most pleasing and beautiful wonders using Fibonacci numbers and, thus, the golden ratio—for example, the pattern of leaves of some succulents (e.g., various subspecies of aloe), seeds in a sunflower, the progressive growth chambers of a chambered nautilus, pinecones, the surfaces of pineapples, and on and on (Figure 8–8).

The message: If for no other purpose than aesthetics in design, engineers should know about the golden ratio.

But now let's use ratios to solve an engineering problem that, at first glance, doesn't seem solvable with the equation at hand.

Illustrative Example 8-3: Using Ratios to Solve a Problem

A technician at the company for which you work as an engineer gives you a printout of Intensity versus Diffraction Angle 2θ that shows five diffraction peaks for an unknown crystalline solid, but you are not told what wavelength λ x-radiation was used. You are asked to tell the sender whether the unknown substance has a face-centered cubic (FCC) or a body-centered cubic (BCC) crystal structure.

(a)   (b)

Figure 8-8 The golden ratio shows up in Nature as Fibonacci numbers that seem to dictate how plants and some mollusks grow, such as (a) the seeds of a sunflower, as a helianthus whorl, and (b) the progressive growth chambers of the chambered nautilus. (*Sources:* WikiMedia, [a] originally contributed by L. Shymlal on 20 June 2006 and [b] originally contributed by Chris 73 on 14 August 2006, both used under Creative Commons free license without requiring permission.)

Looking up Bragg's law for x-ray diffraction, you find it to be

$$n\lambda = 2d_{hkl} \sin \theta$$

within which d_{hkl} is the distance between adjacent parallel planes with Miller indices (hkl).[4] Even assuming all first-order diffraction, where $n = 1$, it appears that without the wavelength λ and some knowledge about the crystalline solid so that you knew d_{hkl} for each 2θ diffraction peak, there are too many unknowns. But, in fact, you have the five conditions that produced diffraction peaks at 2θ values of $2\theta_1 = 40.0°$ ($\theta_1 = 20.0°$), $2\theta_2 = 58.0°$ ($\theta_2 = 29.0°$), $2\theta_3 = 73°$ ($\theta_3 = 36.5°$), $2\theta_4 = 87°$ ($\theta_4 = 43.5°$), and $2\theta_5 = 101°$ ($\theta_5 = 55.5°$). Thus, you can write a set of five equations:

$$n\lambda = 2d_{h1k1l1} \sin \theta_1$$
$$n\lambda = 2d_{h2k2l2} \sin \theta_2$$
$$n\lambda = 2d_{h3k3l3} \sin \theta_3$$
$$n\lambda = 2d_{h4k4l4} \sin \theta_4$$
$$n\lambda = 2d_{h5k5l5} \sin \theta_5$$

Because the same radiation was used to generate all five peaks, λ, and thus $n\lambda$, are constant. Hence, as you should remember, the basic principle in mathematics that "things that are equal to the same thing are equal to one another" allows the right side of each of the preceding equations to be equated to one another as:

$$d_{h1k1l1} \sin \theta_1 = d_{h2k2l2} \sin \theta_2 = 2d_{h3k3l3} \sin \theta_3 = d_{h4k4l4} \sin \theta_4 = d_{h5k5l5} \sin \theta_5$$

as even the 2s cancel.

You also know, from the Internet (or from a course in basic materials) that $d_{h1k1l1} = a_o/(h_1^2 + k_1^2 + l_1^2)^{1/2}$, etc., for each peak; in which a_o, the lattice parameter, is also constant for the unknown crystal. Thus, it is possible to relate ratios of the different $\sin \theta$ values

---

[4]Miller indices are a system of notation for points, directions, and planes in a crystal. Good explanations of how Miller indices are determined for a point, a direction, or a plane in a crystal are available on the Internet.

for subsequent pairs of diffraction peaks, 2 to 1, 3 to 2, 4 to 3, and 5 to 4, to ratios of subsequent values of $(h^2 + k^2 = l^2)^{1/2}$, so that, for example:

$$\sin \theta_2 / \sin \theta_1 = (h_2^2 + k_2^2 + l_2^2)^{1/2} / (h_1^2 + k_1^2 + l_1^2)^{1/2}, \text{ etc.}$$

The values of these ratios for the five peaks are:

$$\sin \theta_2 / \sin \theta_1 = \sin 29° / \sin 20° = 0.4848/0.3420 = \textbf{1.418}$$
$$\sin \theta_3 / \sin \theta_2 = \sin 36.5° / \sin 29° = 0.5948/0.4848 = \textbf{1.227}$$
$$\sin \theta_4 / \sin \theta_3 = \sin 43.5° / \sin 36.5° = 0.6884/0.5948 = \textbf{1.157}$$
$$\sin \theta_5 / \sin \theta_4 = \sin 50.5° / \sin 43.5° = 0.7716/0.6884 = \textbf{1.121}$$

The rule for diffraction from an FCC crystal lattice is that h, k, and l must all be even or all be odd, while the rule for a BCC crystal lattice is that $h + k + l$ must be even (where 0 is taken to be even). Thus, from among all possible values of $(h^2 + k^2 + l^2)^{1/2}$, the planes that would produce the peaks for an FCC versus a BCC crystal, alone with the ratios of $(h^2 + k^2 = l^2)^{1/2}$, follow:

Possible *hkl* Value	FCC	BCC
√1 for (100)	—	—
√2 for (110)	—	1st
√3 for (111)	1st	—
√4 for (200)	2nd, ratio = 1.155	2nd, ratio = **1.414**
√5 for (210)	—	—
√6 for (211)	—	3rd, ratio = **1.255**
√7 ... none	—	—
√8 for (220)	3rd, ratio = 1.414	4th, ratio = **1.155**
√9 for (221) or (300)	—	—
√10 for (310)	—	5th, ratio = **1.118**
√11 for (311)	4th, ratio = 1.173	—
√12 for (222)	5th, ratio = 1.044	—

Thus, with matching of the ratios for the first five peaks here and previously, the unknown solid has a BCC crystal structure.

If you knew the wavelength λ of x-radiation used, you could go on to use Bragg's law to find a value of d_{hkl} at one of the peaks—say, d_{110} from the equation $d_{h1k1l1} = a_o/(h_1^2 + k_1^2 + l_1^2)^{1/2} = a_o/(1^2 + 1^2 + 1^2)^{1/2} = a_o/\sqrt{3}$. You could then calculate a_o using the rearranged equation $a_o = d_{hkl}(h^2 + k^2 + l^2)^{1/2}$. With a value for a_o and knowing the crystal structure is BCC, you could find the unknown material from tabulated data available in various references on crystallography.

The technique of using ratios works when you are faced with a problem for which you have a mathematical equation that can be applied to a set of data to generate a set of equations. If these equations are each equal to the same constant (on one side of the equation), they are equal to one another, and the approach of taking ratios will work.

In Summary

Similarity is useful for solving some problems, whether that similarity is geometric similarity alone, or kinematic similarity, or dynamic similarity (in which there must be both geometric and kinematic similarity). While similarity involves things that are proportionate by ratios, there are problems for which *ratios* can be used without the requirement for or existence of similarity. A particularly fascinating ratio is the mathematical constant found in the golden ratio. This ratio has particular value in engineering, as it does in art, architecture, and Nature, for aesthetics. Ratios can also be used to solve certain types of engineering problems when two or more things are equal to the same thing but somewhere else on a scale (e.g., 2θ angles in x-ray diffraction).

CHAPTER 9

Using Indexes or Indices

YOU GET A sense at the checkout counter of your local supermarket that prices for food are going up faster than your income. You're running through your weekly allotment for food earlier in the week than ever. You're probably right. But since you don't buy exactly the same items each week, how could you be sure? Then you hear on the 6 p.m. news that the CPI rose 3 percent over the past six months, so that the average family of four is spending more on necessities now than at any time in the past decade.

Googling "CPI" on your iPad, you find that the Consumer Price Index measures changes in the price level of consumer goods and services purchased by households. The CPI is defined by the U.S. Bureau of Labor Statistics as "a measure of the average change over time in the prices paid by urban consumers for a market basket of goods and services." It's a statistical estimate for a representative set of items whose prices are collected periodically. It allows quantitative comparison of an ill-defined collection or combination of important things. And, while it fluctuates over short periods of time (e.g., year to year), it goes up ceaselessly over the long haul (Figure 9–1).

In solving problems, engineers usually need to quantify things, as physical parameters in mathematical equations and their solutions or as inputs to and outputs of processes. For many things, quantification can be precise, as the parameters, inputs, or outputs are straightforward and easily defined and measured. Examples include values for the mass of an object, the force on an object, the yield strength of a metal, the temperature of a process, the voltage, current, and travel speed for arc welding, or the composition of a chemical product. For some things, however, quantification can be very difficult for any number of reasons, two of the most common being (1) the complexity of a property that is really the combination of more basic properties (e.g., machinability, formability, weldability) and (2) the ill-defined and/or intangible nature of the parameter (e.g., tonality of a musical instrument or quality of a weld). For problems involving either type of factors, characteristics, or qualities, a useful technique is to use *indexes* or *indices*.

The dictionary[1] defines the noun *index* as "something that serves to guide, point out, or otherwise facilitate reference, especially." Acceptable plural forms are *indexes* and *indices*. As they apply to engineering problem-solving, *indexes* or *indices* tend to be used in one of four general ways, as follows.

First, indexes or indices are used *to relate a readily quantifiable property of one material (for example) to another,* typically as a fractional, decimal, or percentage multiplier. Properties for which this is done tend to be those for which the units are less familiar, so a physical sense of the property is more difficult for an engineer to have or develop.

The best example is the way the electrical conductivity of a metal or alloy is related to the familiar example (and reference point) of pure copper. Values of electrical conductivity can be found in units of Ω^{-1}-m^{-1} (kgs) or Ω^{-1}-cm^{-1} (cgs) in many readily available references, but the value for any particular metal or alloy means very little to most engineers on an absolute scale or basis. The more meaningful way to express the electrical conductivity of a metal or an alloy is often the *IACS index.* The International Annealed Copper Standard is based on the *electrical resistivity* of pure (e.g., high-conductivity, oxygen-free, OFHC, copper, or electrolytic tough-pitch, ETP, copper) in a fully annealed (soft, residual stress–free) condition, measured at 20°C as 1.7241×10^{-6} Ω-cm. Electrical resistivity ρ_{el}

---

[1] Definitions are from Houghton Mifflin's online dictionary at www.thefreedictionary.com/Houghton.

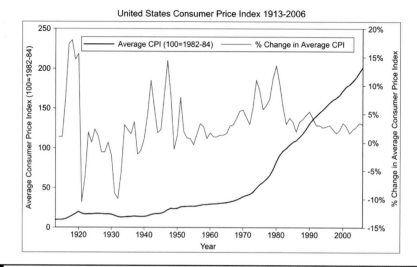

Figure 9-1 The Consumer Price Index (referenced to 1982–84 = 100) from 1913 to 2003. (*Source:* U.S. Bureau of Labor Statistics; no permission required; also Wikimedia Commons, originally contributed by Andrew pmk on 16 June 2003 under Creative Commons Atttribution Share-Alike 3.0 license.*)*

is used instead of electrical conductivity σ_{el}, as it is far easier to measure resistance (in Ohm's law $R = E/I$, resistance in ohms Ω is obtained from voltage in volts V over current in amperes A)[2]. In the IACS index, a value of 100 is assigned to the pure copper standard, the value of conductivity for other metals or alloys being expressed as a percentage of this standard (Table 9–1).

Second among the four ways in which indexes or indices are used is *to give semiquantitative values to allow rank-ordering of complex behaviors, characteristics, or properties,* with good examples being some ill-defined but recognizable measure of quality (e.g., tone or tonality of a musical instrument or quality of a weld) and complex properties important to manufacturing processes (as such things as formability, for example, arise from some undefined combination of easily measured material properties such as hardness, strength, and ductility, at least).

A very good example is the *machinability index* that relates the ease with which a metal or alloy can be machined (e.g., by milling, turning, or drilling) to an acceptable surface finish relative to AISI 1212 machinable-grade carbon steel (at 100 percent), expressed as a percentage.[3] Table 9–2 summarizes some values for the machinability index for a variety of steels in various categories. Other manufacturing process complex combined properties of importance for which indices are used include formability, castability, moldability, weldability, brazeability, and solderability (Ref. Messler).

Third among the four ways in which indexes or indices are used is to assess the relative effectiveness of a mechanical, electrical, thermal, optical, or material phenomenon that actually involves several different factors or properties for which there may be irreconcilable or difficult-to-interpret units.

---

[2]The more common symbols for electrical resistivity and electrical conductivity are ρ and σ, but these can be confused for density and stress or strength without the "el" subscript used here. Electrical resistivity and electrical conductivity are reciprocals of one another, i.e., $\rho_{el} = 1/\sigma_{el}$ and $\sigma_{el} = 1/\rho_{el}$, as are their units of Ω-m and Ω^{-1}-m^{-1}, for example. Resistance $R = \rho_{el}\,(L/A)$, for a physical conductor in which L is the conductor's length and A is its cross-sectional area, while ρ_{el} is the material's electrical resistivity, as an inherent hindrance to conduction by electrons.

[3]American Iron and Steel Institute, AISI-designated 1212 is a standard resulfurized and rephosphorized grade carbon steel containing 0.13 percent (max.) C, 0.70–1.00 percent Mn, 0.07–0.12 percent P, and 0.16–0.23 percent S. It was specially developed (many years ago) as a highly machinable grade carbon steel that, by allowing easy chip formation/breakage, provides high-quality as-machined (versus ground) surface finishes.

TABLE 9-1 Values of Electrical Conductivity of Some Important Metals and Alloys in Engineering Using the IACS Index

Material	IACS Conductivity
Silver (pure)	105%
Copper (pure)	100%
Gold (pure)	69%
Aluminum (pure)	60%
Tungsten (pure)	30%
Brass (Cu-Zn alloys)	28–32%
Zinc (pure)	27%
Nickel (pure)	23%
Iron (pure)	17%
Bronze (Cu-Sn alloys)	7–15%
Titanium (pure)	4%
Steel (plain carbon)	3–15%
Steel (stainless)	2.4%

TABLE 9-2 Values of the Machinability Index for Some Important Engineering Alloys

	Index*
Carbon steels (annealed)	100%
AISI 1212	78%
AISI 1018	70%
AISI 1030	64%
AISI 1040	54%
AISI 1050	
Low-alloy steels (annealed)	72%
AISI 4130	66%
AISI 4140	57%
AISI 4340	
Stainless steels (annealed)	45%
AISI 304 austenitic	45%
AISI 316 austenitic	110%
AISI 410 martensitic	54%
AISI 430 ferritic	
Tool steels (annealed)	27–42%
Cast irons (annealed gray, ductile, or nodular)	27–40%
Wrought Al alloys	360%
Cast Al alloys	450%
Wrought Mg alloys	480%
Cast Mg alloys	480%
Wrought Ti alloys (annealed)	50-70%

*Relative to AISI 1212 at 100%.

A good example is the thermal shock resistance *TSR* parameter or *index,* used to assess the relative resistance to cracking or complete fracture a material offers to a rapid change (usually a drop) in temperature (e.g., a hot glass casserole dish run under cool water). This behavior is important to know for certain service applications (e.g., space shuttle tiles during reentry into the atmosphere from outer space) or processes (e.g., quenching steel parts during heat treatment or rapid heating and/or cooling during arc welding). The TSR index includes the effects of the fundamental material properties of fracture strength σ_f (in MPa or 10^6 N-m/kg^2 or lbm/in^2 or psi/ksi), linear thermal expansion coefficient

TABLE 9-3 Calculated Values of the TSR Parameter or Index for Some Important Engineering Materials					
Material	σ (MPa)	*E* (GPa)	*k* (W/m-K)	α_l ($\times 10^{-6}$°C^{-1})	TSR (kW/M)
Steel (1020 annealed)	395	207	51.9	11.7	8.50
Steel (4340 Q&T)	1760	207	44.6	12.3	30.8
Stainless (304 annealed)	515	193	16.2	17.3	2.50
Gray cast iron	276	138	46.0	11.4	8.10
Al alloy (7075 aged)	572	71	130	23.4	45
Brass (70Cu-30Zn)	330	110	120	19.9	18.1
Ti-6 Al-4V (annealed)	830	114	6.7	8.6	5.7
Ni alloy 625 (annealed)	930	207	9.8	12.8	3.4
Concrete (compression)	337	31	1.5	11.8	1.4
Glass (soda-lime)	69	69	1.7	9.0	0.19
PyroCeram	250	120	3.3	6.5	1.1
Aluminum oxide	420	380	39	7.4	5.8
Silicon carbide	530	345	80	2.9	42
Epoxy	60	2.4	0.19	102	0.05
Polyvinyl chloride	42	3.3	0.18	135	0.02
Rubber (butadiene)	15	0.0034	0.25	235	4.7
Fir, parallel to grain	108	12.2	0.14	4.5	0.28
Oak, parallel to grain	112	12.7	0.18	5.3	0.30

α_l (typically as 10^{-6}°C^{-1} or K^{-1} or 10^{-6}°F^{-1}),[4] thermal conductivity k (in W/m-K or Btu/ft-h-°F), and modulus of elasticity E (in GPa or 10^9 N-m/kg^2 or psi/ksi) expressed as

$$TSR \sim \sigma_f \, \alpha_l / k \, E$$

You should note that the units for these index parameters do not necessarily result in any meaningful or, at least, physically understandable units, as

$$[(\text{N-m/kg}^2)(\text{K}^{-1})]/[(\text{W/m-K})(\text{N-m/kg}^2)] \Rightarrow \text{W/m}$$

although, here for TSR, watts (W) per m denotes how fast the heat energy in a physical object changes (e.g., drops) over a distance, which clearly would affect the severity expansion/contraction-induced stresses would develop (with lower stresses for greater spreading of heat). A higher value of TSR index indicates a greater resistance to thermal shock, but not necessarily with any absolute meaning to the relative magnitudes of the index for the different materials.[5]

Table 9–3 summarizes some calculated values for the TSR index for a variety of important/common materials used in engineering applications. It is noteworthy that the values for the TSR index for polymers (including rubbers and woods, as polymer-polymer composites) are much lower, compared to other materials, than one would expect intuitively. One must think about severe temperature changes, compared to what are found in Nature, which do cause polymers to experience damage by crazing or cracking.

Illustrative Example 9–1: Using TSR Index

A friend tells you there is no better material for a casserole dish than PyroCeram.[6] You tell your friend that your mother gave you Pyrex glass[7] casserole dishes as a gift, saying "Pyrex is the best when it comes to not breaking when it comes from a hot oven." You decide to put an end to the debate by calculating the thermal shock resistance using the TSR parameter or index for both materials to see which is really better.
From the Internet, you get the needed properties for the two materials as:

	PyroCeram	Pyrex
σ_f (MPa)	250	69
E (GPa or 10^3MPa)	120	70
k (W/m-K)	3.3	1.4
α_l ($\times 10^{-6}$°C^{-1})	6.5	3.3

---

[4] In fact, the effect of thermal expansion/contraction can be represented by the linear coefficient of thermal expansion α_l or the volume coefficient of thermal expansion α_v.

[5] The logicality of the arrangement of the four material-property parameters in the TSR index that allows one to reconstruct it (see Chapter 4), if necessary, is that the resistance to damage from thermal shock would go up as the fracture strength of the material goes up and as heat spreads to lessen temperature gradients (and thermally induced stresses) due to a higher thermal conductivity. Thus, both would appear in the numerator—for their direct effect. The more a material expands, the greater the tendency for that material to crack from thermally induced stresses (i.e., the lower the resistance to shock), which puts the coefficient of thermal expansion in the denominator, for its inverse effect. Finally, the stiffer a material (i.e., the higher E is), the more likely it is to crack, so E goes in the denominator for its inverse effect.

[6] PyroCeram was developed and trademarked by Corning Glass in the 1950s. It is produced by the controlled crystallization of a special glass formulation to create what is generically known as a *glass-ceramic* material.

[7] Pyrex was developed by Corning Glass in 1915. Originally a low-CTE (coefficient of thermal expansion) borosilicate glass, in the 1940s some varieties where made from tempered soda-lime glass.

Plugging these basic values (without concern for orders of magnitude, because they are all the same) into the equation for TSR ~ $\sigma_f k / E\alpha_l$ and "chugging" gives:

(250)(3.3)/(120)(6.5)	(69)((1.4)/(70)(3.3)
~1.06	~0.39

The 3X-higher value of TSR index for PyroCeram gives credence to modern materials development, rather than risk saying that mothers don't know best!

Last but not least among the four ways in which indexes or indices can be used by engineers as problem-solving techniques is to help assess the performance of a structural component during design in terms of the performance-limiting material properties. This approach, widely and popularly advanced by the University of Cambridge's Michael F. Ashby in his books and software on material selection in mechanical design (Ref. Ashby), involves partitioning an overall function that expresses the needed or desired performance of the component (e.g., to resist deflection as a cantilevered beam while minimizing weight or maximizing the amount of energy that can be safely stored in a flywheel while limiting the volume occupied by the flywheel) into subfunctions that contain geometric parameters only, functional parameters only (generally also including number-constants), and material parameters or properties only. The material subfunction represents what is known as the *material performance index,* or MPI. By maximizing this index, the mass/weight, volume, energy, cost, or even environmental impact can be optimized. Familiar examples of important MPIs are stiffness to weight (or specific stiffness), as E/ρ, and strength to weight (or specific weight), as σ_y/ρ or σ_f/ρ, in which ρ is the density of the material.

The power of the use of a material performance index (or, for designs with multiple limiting properties, material performance indices) is that materials with equal values of the derived MPI are structurally comparable *for those properties included in the MPI.* Ashby offers full-spectrum plots of one property in the MPI against the other (e.g., Young's modulus versus density) in what are called *material selection charts.* From such charts, using design guidelines for similar designs (see Chapter 16, "Reverse Engineering"), materials that are comparable or superior to some material known to have been used successfully can be located along or above the line.[8] Everyone knows, for example, that tennis rackets can be made from wood, glass fiber–epoxy composite, or graphite fiber–epoxy, while the support structure for roller coasters can be made of wood, steel, or concrete.

Table 9–4 offers a sampling of MPIs for various design functions and constraints. Figure 9–2 shows an exemplary material selection chart.

Illustrative Example 9–2: Deriving and Using a Material Performance Index

The main fore-to-aft support beams beneath the bed of a flatbed trailer truck must resist deflection under the maximum load allowed by the Department of Transportation so as not to fail by buckling. As beams supported at one end (i.e., the hitch) by a pin or hinge and at the other end by a roller (i.e., the rear-wheel assembly), the equation to predict elastic deflection δ is

$$\delta = FL^3/48EI$$

in which F is the load (assumed—but not usually—concentrated at the midspan of the trailer and beams), L is the length of the beam (here, beams), E is the modulus of elasticity

---

[8]Ashby's book, *Materials Selection in Mechanical Design,* gives other ways than using reverse engineering to arrive at an appropriate design guideline's magnitude for a given combination of properties.

TABLE 9-4 Values for the Material Performance Index for Some (of Many) Design Functions and Constraints*

Stiffness- or Strength-Limited at Minimum Mass:	Maximize	
Tensile strut or tie	E/ρ	σ_f/ρ
Shaft (in torsion)[1]	$G^{1/2}/\rho$	$\sigma_f^{2/3}/\rho$
Beam (in bending)	$E^{1/2}/\rho$	$\sigma_f^{2/3}/\rho$
Column (in compression)	$E^{1/2}/\rho$	σ_f/ρ
Plate[2]	$E^{1/3}/\rho$	$\sigma_f^{1/3}/\rho$
Flywheels[3]	—	σ_f/ρ
Damage Tolerance-Limited, Independent of Mass:	**Maximize**	
Ties	K_{IC} and σ_f	
Shafts	K_{IC} and σ_f	
Beams	K_{IC} and σ_f	
Pressure vessels		
Thick-walled designed to yield before break	K_{IC}/σ_f	
Thin-walled designed to leak before break	K_{IC}^2/σ_f	

*Values were selectively extracted from values published in Michael F. Ashby's *Materials Selection in Mechanical Design*, 3rd edition, Elsevier, 2005; used with permission.
[1]Length and cross-sectional shape specified; cross-sectional area free.
[2]Length and width specified; thickness free.
[3]For maximum energy storage per unit mass.

of the material used to fabricate the beam (here, beams), and I is the moment of inertia for the beam (here, each beam).

To minimize fuel consumption—and maximize the net weight of cargo that can be carried—the weight or mass of trailers (and, thus, all of their structural elements) must be minimized.

Real support beams for flatbeds, in order to minimize weight, have an I-shaped cross section for greater stiffness against bending and tend to vary in cross-sectional area from fore to aft, being of maximum cross section near the midspan. For simplicity in this example, the beam cross section will be considered to be square and solid and of uniform cross-sectional area along its entire length. Further, two identical beams will be assumed, one running just inside each edge of the trailer. Flatbed trailers actually are not flat but, rather, have an upward bow or hump so as to cause allowable deflection to remove the bow or hump but never go past flat or horizontal. Thus, in an actual design, one would calculate how much deflection δ would occur under maximum allowable loading and build that value into the height h of the bow, with some factor for safety, of course.

To find the material performance index (MPI) that will maximize resistance to deflection due to bending at minimum weight or mass, two equations must be combined. First, the full equation dictating performance of the beam, including an expression for the moment of inertia, must be written as:

$$\delta = FL^3/48EI, \text{ where } I = a^4/12 \text{ for a solid square of side } a$$

or

$$\delta = (FL^3)/(48E)(a^4/12)$$

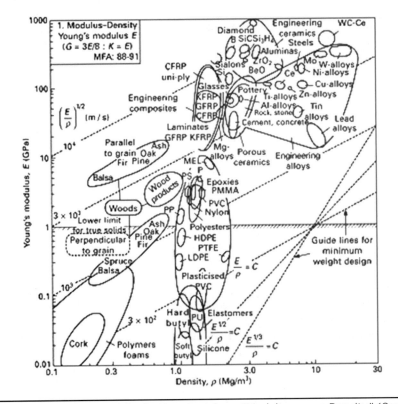

Figure 9-2 An example of a material selection chart, here, "Modulus versus Density." (*Source:* Michael F. Ashby, *Material Selection in Mechanical Design*, 3rd edition, Elsevier, 2005; used with permission.)

which reduces to

$$\delta = FL^3/4Ea^4$$

Next, the equation for the mass m, as density times volume, of the two *beams* must be written as:

$$m = \rho a^2 L(2) = 2\rho a^2 L$$

The common geometric parameter (or physical dimension) between the performance equation and the equation for mass over which you, as the designer, have control is the length of the side of the square beam a, as the length L of the beam (here, beams) is fixed by the length of the trailer (which tends to be fairly standard at around 53 ft in the United States).

Using the performance equation to solve for a, a is written as:

$$a = (FL^3/4E\delta)^{1/4}$$

Substituting this value for a into the equation for the mass m gives:

$$m = (2)\rho[(FL^3/4E\delta)^{1/4}]^2(L) = 2\rho[(FL^3/4E\delta)^{1/2}]L$$

or

$$m = \rho F^{1/2} L^{3/2}/2E^{1/2}\delta^{1/2}$$

Separating (or partitioning) this equation into parts—or subfunctions—that contain only geometric parameters (here, only L), only functional parameters, and any number of constants (here, F, δ, and 2), and only material parameters (here, ρ and E), gives:

$$m = [L^{5/2}][F^{1/2}/2\delta^{1/2}][\rho/E^{1/2}]$$

The last subfunction represents the material parameters only. To minimize m, the value of $\rho/E^{1/2}$ must be minimized, which means that the reciprocal, $E^{1/2}/\rho$, must be maximized. The MPI for this stiffness-limited design of support beams is $E^{1/2}/\rho$. This MPI is consistent with the expression given in Table 9–4.

With the MPI derived, one can go on, using Ashby's methodology (Ref. Ashby), to employ a material selection chart that plots the full-spectrum values for one property versus another in an MPI, here, E versus ρ, that is, Young's modulus versus density (Figure 9–2). By drawing a line parallel to the dashed design guide lines with slope $E^{1/2}/\rho$ that passes through a material known to have been successfully used in beams, for example, "Steels" (near the upper-right corner), other materials with a value of $E^{1/2}/\rho$ that are comparable or superior fall on or above the line, respectively. Here, viable candidates include certain Al alloys and CFRP (carbon-fiber-reinforced polymer) composites. Notice that hardwoods, such as ash and oak are nearly comparable, which clarifies why they were, in fact, used in trucks and trailers in the 1920s.

The decision of which to use depends on other factors, such as cost, availability, ease of fabrication into beams with more complex cross sections (e.g., I-beams) and with a variable cross-sectional area along their length, and so on.

In Summary

It is sometimes useful—and other times it is necessary—in engineering, as part of solving problems, to compare factors, characteristics, behaviors, properties, or qualities. This can sometimes be done more (if not most) efficiently, and for hard-to-quantify behaviors or complex manufacturing-oriented properties (e.g., machinability) can only be done, using *indexes* or *indices*. Four general situations in which indexes or indices are useful are: (1) to relate one measurable/quantifiable parameter (e.g., property) to another when the units of the parameter are unfamiliar and/or hard to grasp in a physical sense (e.g., electrical conductivity); (2) to give semiquantitative values to allow rank-ordering of complex behaviors, characteristics, or properties (e.g., machinablity); (3) to assess the relative effectiveness of a mechanical, electrical, thermal, optical, or material phenomenon that actually involves several different factors or properties for which there are no reconcilable or meaningful units (e.g., TSR); and/or (4) to help assess the performance of a structural component during design in terms of the performance-limiting material properties (as a material performance index).

Suggested Resources

Ashby, Michael F., *Materials Selection in Mechanical Design*, 3rd edition, Elsevier, Burlington, MA and Oxford, UK, 2005.

Messler, Robert W., Jr., *Essence of Materials for Engineers*, Jones & Bartlett Learning, Sudbury, MA, 2010.

CHAPTER 10

Scaling

APROPOS TO MATHEMATICAL approaches or techniques for solving problems in engineering, the term *scaling* is defined as it applies to euclidean geometry, as "a linear transformation that enlarges (increases) or shrinks (diminishes) objects by a [scale] factor that is the same in all directions."[1] The result of such *uniform scaling* or *isotropic scaling,* of course, is a geometrically similar and congruent version (e.g., model) of the original (e.g., prototype) (see Chapter 8).[2] It is possible in engineering to have nonuniform scaling (or *anisotropic scaling* or *inhomogeneous dilation*) for which the scaling factor is different in some directions than in others. Nonuniform scaling changes the shape of the object between the prototype (original) and the model (modified version), so geometric similarity is lost. A familiar example is the distortion that results when a motion picture filmed for the theater's big screen is shown on television. Because the aspect ratios (i.e., width to height) for the movie screen (really, the film) and for the TV screen are different, reducing the film images for the TV screen distorts, by shortening. (This may be why one of your favorite movie stars seems a little shorter and chubbier in the TV movie than at the theater.) The other commonly seen alternative is to maintain the aspect ratio of the original film and black out or frame the TV screen at the left and right sides, eliminating direction of the picture.

As a problem-solving technique in engineering, *scaling* is useful in two situations:

1. To create a model that is (occasionally) larger or (usually) smaller than the original or planned prototype. These are known as *scale models.*
2. To facilitate variant (scaled) design.

Relative to the first situation, subscale models are often used in engineering to assess the likely behavior of the full-size prototype at a much-reduced cost for the model and for the testing facility. For the results of tests on subscale models to be meaningful to the full-size prototype, however, there must usually be kinematic and, often, dynamic similarity (see Chapter 8).

Scale models are built and test data are collected for many reasons in engineering but the most common reasons, by far, are (1) to check out the feasibility of some new design and/or (2) to assess the likely performance of a design object at an early stage without the expense of building a full-size prototype (see Chapter 20, "Test Models and Model Testing"). Scale model testing is routinely done in the automobile industry (Figure 10–1), in the aerospace industry (Figure 10–2), and in the field of naval architecture for shipbuilding (Figure 10–3) to generate data on aerodynamics or, for the last category, hydrodynamics, for optimized performance and/or fuel economy.

Scale models for testing in wind tunnels (for automobiles or aircraft) or flow tanks (for ships) must be scaled to be kinematically and dynamically similar more than just geometrically similar. The former is to ensure like fluid flow (for streamlining); the latter, to replicate and assess the effects of forces (for structural integrity). For aircraft and ships, which must operate in and use a fluid, testing of scale models in wind tunnels or flow tanks is critical to the assessment of stability and control.

Architects, not just engineers, use scale models. However, more than for assessing performance,

---

[1] Definition is from www.encyclopedia.thefreedictionary.com, a website of the Houghton Mifflin Company.
[2] A scale factor of 1 is allowed and results in a shape that is congruent, besides being similar.

Figure 10-1 Wind tunnel testing of an open-wheel Formula 3 race car, for which there must be kinematic and dynamic similarity beyond geometric similarity by scaling the car's dimensions. Here, the scaling factor is 1, making the test model full size. (*Source:* Photograph from MIRA Ltd., Nuneaton, Warks, U.K., used with permission.)

architects use models to assess aesthetics—that is, the fit of the design to a specific site and/or purpose. Scaling for architecture is usually restricted to geometric scaling to assess geometric similarity for evaluation of style. For civil engineers, however, as the ones responsible for ensuring that an attractive design for a building by an architect will stand and perform structurally, scale models are used for kinematic and dynamic simulation. Civil engineers routinely employ scale models of large buildings, bridges, and dams (Figure 10–4).

There are a variety of physical approaches to assist in scaling either the design for a prototype or an actual object to a model for testing. In both architecture and in engineering, special scaling rulers were commonly used before computer-aided graphics came about. These allowed direct transfer of an original dimension (say, 14 ft 4 in) to a drawing or model for certain scales (1/4 in = 1 ft, 1/16 in = 1 ft, etc.). Other graphic techniques are described in Part 3 of this book.

(*a*)   (*b*)

Figure 10-2 Wind tunnel testing of a 3 percent scale model at NASA's Langley Research Center, Hampton, VA (*a*), of the Boeing-designed X-48B blended-wing body technology demonstrator aircraft at Rogers Dry Lake adjacent to NASA Dryden (*b*). (*Source:* From an article by Sean Smith, NASA Langley, [*a*] contributed by Uwe W on 20 December 2006 and [*b*] contributed by Originalwana on 16 March 2011, both used for free as part of the public domain.)

Figure 10-3 "Sea trial" testing of the largest model ship ever tested at Marintek's Ocean Basin Laboratory in Trondheim, Norway, by Teekay Corporation. (*Source:* Photograph from Marintek Ocean Test Basin, Marintek, Trondheim, Norway, used with permission.)

Illustrative Example 10-1: Using Scaling

You decide to try your hand at building an HO-gauge boxcar like one you saw—and photographed—as it passed through a crossing at which you were stopped in your hometown. You find the specifications for a standard boxcar in the United States on the Internet to be as follows:

Length = 50 ft
Width = 9 ft 6 in
Height = 13 ft $^{13}/_{16}$ in
Side-door height = 12 ft 4 in; width = 10 ft

You also find the scale factor for HO gauge to be 1:87.
    With this information, you calculate the scaled dimensions to the nearest 1/16 in, rounding as required, to be as follows:

Length = 50 ft/87 = 0.547 ft = 6⅞ in
Width = 9 ft 6 in/87 = 1.31 in = 1$^{5}/_{16}$ in

Figure 10-4 Scale model of the ship lift (just to the right of the dam, in the background) and double five-step locks (right of the ship lift and its own waterway, for the Three River Gorge Dam on the Yangtzee River near Wuhan in China. Models like this are often used during the planning stage for such a massive 25-year project. (*Source:* Wikimedia Commons; originally contributed by Shizhao on 30 June 2005 and freely used under Creative Commons Attribution Share-Alike 1.0 license.)

Figure 10-5   A photograph showing a family of Airbus commercial transport airplanes, with clear relationships between the basic configurations that suggest a role of variant design. (*Source:* Photograph provided by Airbus, and used with permission courtesy of that company.)

Height = 13 ft $^{13}/_{16}$ in/87 = 1.80 in = 1$^{13}/_{16}$ in
Door width = 10 ft/.87 = 1.38 in = 1$^{3}/_{8}$ in
Door height = 12 ft 4 in = 1.70 in = 1$^{11}/_{16}$ in

Now you're ready to build your model boxcar.

Scaling is also used, as was stated earlier, in *variant design,* in which the design for a new object, part, structure, or the like, is largely the result of a scaled version of the original. Examples of fairly straightforward variant designs include stretch limousines and sport utility vehicles (SUVs), stretch or otherwise derivative commercial airliners (Figure 10–5), large-size TV screens, and downsized furniture for apartment dwellers in cities. Obviously, there is a need, in each, to ensure that kinematic and dynamic similarity is considered and dealt with. Variant design, involving fairly straightforward geometric scaling, is done with artificial hip, knee, and other joint replacements, so that the replacement joint properly fits the recipient. To keep costs manageable, total customization is impractical.

There are products, such as smaller computers (e.g., from desktop PCs to portable notebooks); smaller, less obtrusive hearing aids (e.g., from those worn outside and behind the ear to in the ear); and larger TV screens (from cathode-ray tubes to LCDs, plasma displays, or projection types) in which one or more of new technology, new materials, and new processing methods are needed, beyond simple geometric scaling. However, even for such products, geometric scaling is still involved.

In Summary

Scaling is a useful technique for solving certain engineering problems, provided the scaling considers kinematic and/or dynamic similarity beyond simple geometric similarity when appropriate. Scaling is most popular for the creation of smaller models (i.e., scale models) of a large original or prototype to allow testing of key aspects of, for example, proper operation of a mechanism, proper aerodynamics (for streamlining), and for assessment of aesthetics. For some scaling, up or down, as is done in variant design, new technology(ies), new material(s), or new process(es) of manufacture are needed to allow geometric scaling.

CHAPTER 11

Sensitivity Analysis (of Parameters)

IN CHAPTER 3, on using mathematical equations to solve problems, it was pointed out that it is important for an engineer to recognize and consider the *form* of the equation. The reasons given were—and are: (1) to help understand the physical significance of each parameter in the equation, as an input to the equation, using units; (2) to help understand the physical significance—and operative physics—of the overall equation due to the interaction of the input parameters (and their units); and (3) to help consider the physical significance of the solution to the equation (as an output or outcome). Different forms of mathematical equations express which parameter(s), as input factors or properties, affect the solution, as an output or outcome *and* to what degree (e.g., power) each does so. This latter aspect of the equation involves what is being referred to here as *sensitivity analysis.*

Sensitivity analysis, quite simply, considers the effects of changes to the independent variables of an equation on the dependent variable of that equation (i.e., the solution to the equation).[1] The reason engineers need to (or should) conduct sensitivity analysis is to be aware of which input has the greatest effect on an output. In some cases, the need or desire is to have the greatest effect on the output that is possible or practical (often at the lowest cost) through manipulation of an input or of inputs. Other times, the need or desire is to minimally affect the outcome due to a change in one or more of the inputs. An example of the former situation is to most efficaciously affect a desired structural property, such as the stiffness of a beam to resist intolerable elastic deflection. An example of the latter situation is to minimally change the quality and/or quantity of a desired output to a process (e.g., heat-treated strength) due to some unintentional error in or fluctuation of an input parameter (e.g., temperature or time at temperature). The former seeks maximum responsiveness, while the latter seeks maximum robustness (i.e., resistance) to change.

The following aspects or characteristics of an equation need to be—or should be—considered during sensitivity analysis:

1. Additive (or subtractive) terms in the equation, as these represent factors in a process or behaviors in a material, structure, or device that increase (or decrease) the magnitude of an outcome accordingly, i.e., having reinforcing or nullifying effects.
2. Multiplicative terms in the equation, as these represent factors in a process or behaviors in a material, structure, or device that have a pronounced effect on the magnitude of an outcome (compared to additive or subtractive terms), with two possibilities:
   a. Multiplication of terms with values greater than unity, as these can drastically magnify the outcome (as with synergistic affects).

---

[1] Sensitivity analysis can also be applied to processes, to see the direction and magnitude of an input variable or operating parameter on the output material or process. Used this way, sensitivity analysis usually seeks consistency and stability in the outcome, known as *process robustness.* In this context, one is specifically interested in responses, as discussed in Chapter 12, "Response Curves and Surfaces").

    b. Multiplication of terms with values less than unity (i.e., fractions, decimals, or percentages), as these can drastically diminish the outcome. (A similar effect is seen for terms that appear in the numerator versus in the denominator of an equation,)

3. Terms raised to a power (i.e., exponential terms), as these terms (and corresponding factors in a process or behaviors in a material, structure, or device) have the greatest effect for powers greater than 1 in a numerator or less than 1 in a denominator, and vice versa. (A term raised to a power results in a greater rate of change to the outcome for a given change to that input, i.e., the slope of a plot of the dependent variable (outcome) as a function of the independent variable (input) is greater in magnitude, but so does a larger value for the slope m in a linear equation of the general form $y = mx + b$.) This is discussed in Chapter 3.

The best way to help you with these considerations of, as guidelines to, conducting sensitivity analysis is through an illustrative example of each.

Illustrative Example 11–1: Additive Factors

Electrons are impeded in their movement through a metal due to scattering events that take place wherever and whenever there is a disruption to (perturbation of) the otherwise regular and periodic electric field associated with each atom in the regular crystalline array. Impeded flow of electrons due to such scattering is what gives rise to the metal's resistivity ρ_{el}. Three factors concerning a crystalline pure metal affect its resistivity, causing the resistivity to increase. These three factors are: (1) the effect of temperature above absolute zero (0 K), as increased thermal vibration causes the electric field of a stationary atom to spread and oscillate, increasing scattering and, thus, ρ_{th}; (2) the effect of any foreign atom on the normal electric field due to the "impurity's" different field,[2] as ρ_i; and (3) the effect of line imperfections known as *dislocations* on the strain field and electric field in the vicinity of the dislocation, as ρ_d.[3]

The equation for the total electrical resistivity of a metal, as Matthiessen's rule, is:

$$\rho_{total} = \rho_{th} + \rho_i + \rho_d$$

For pure copper (Cu) at room temperature (RT = 20°C), the value of $\rho_{th} = 1.72 \times 10^{-8}$ Ω-m. Substituting nickel (Ni) atoms for just 1 percent of all of the Cu atoms (i.e., 1 at.% Ni), leads to a value for $\rho_i \sim 1.4 \times 10^{-8}$ Ω-m. Deforming the pure Cu by modest cold work (as in drawing into wire) leads to a value of $\rho_d \sim 0.4 \times 10^{-8}$ Ω-m. Thus, the value of ρ_{total} is:

$$\rho_{total} = 1.72 \times 10^{-8} + 1.4 \times 10^{-8} + 0.4 \times 10^{-8} \text{ Ω-m} = 3.52 \times 10^{-8} \text{ Ω-m}$$

Since the equation for $\rho_i = Ac_i(1-c_i)$, in which A is a composition-independent constant for a particular host and impurity (here, Cu and Ni, respectively), and c_i is the composition of the added "impurity," for small concentrations of "impurity," doubling the amount of added Ni would have a much more significant effect on overall resistivity at RT than doubling the amount of cold work. Thus, "impurity" content is the most significant factor in resistivity, unless temperature is increased significantly.

[2] In alloys, foreign atoms are intentionally added to alter properties of the host metal, most notably by the fact that, being larger or smaller than the host atom, they introduce a strain field that interacts with other strain fields (such as around dislocations) and they disrupt the electric field to cause increased scattering. Impurity atoms, on the other hand, are not removed from raw ingredients or do not come into an alloy accidentally, although they too act as solid-solution strengtheners, albeit often with inordinate adverse effects on the ductility or electrical resistivity and conductivity.

[3] Dislocations allow metals to deform by a process of slip, in which layers of atoms in close-packed planes move over one another in close-packed directions. The number (actually density per unit volume) of dislocations in a metal or alloy is increased many orders of magnitude by cold working (i.e., deforming) the metal near room temperature, causing dislocations to interact with one another through their strain fields.

Illustrative Example 11-2: Multiplication by Factors Greater Than Unity

The elastic deflection δ of a beam (such as the one described in Illustrative Example 9-2), under a load P applied at the midspan, is given by the equation:

$$\delta = PL^3/48\,EI$$

in which L is the length of the beam (often fixed, as in Illustrative Example 9-2, by the length of the flatbed trailer), E is the modulus of elasticity of the beam material, and I is the moment of inertia for the beam as influenced by its cross-sectional shape (taken to be a solid square in the example).

For a given load P to be applied at the midspan of a beam of length L, the resistance to elastic deflection δ comes from the stiffness of the beam as the product of E and I, that is, EI. Since each term is linear (i.e., to the power of 1), an increase in one will have no more effect than an increase in the other. However, the value of Young's modulus E for common engineering alloys doesn't differ much—for example, for steel E = 207 GPa, while for Al alloy E = 70 GPa, a factor of 3X. On the other hand, the value of I depends on how mass in a cross section is distributed around the centroid (center of mass), with shapes such as I-beams having a value of I four to six times the value of I for a solid square beam.

Thus, a change in I, via shape factor, can have a much greater effect on reducing elastic deflection than a change in alloys (i.e., E).[4]

Illustrative Example 11-3: Multiplication of Fractions

The probability of tossing a coin and getting it to land heads up (i.e., "heads") is 1 out of 2 or 1/2. The probability of a coin landing heads up on the second toss is also 1/2. Likewise for each individual toss, the chance of getting "heads" is 1/2.

However, the odds of tossing five "heads" in a row (i.e., consecutively) is

$$P_{overall} = P_1 P_2 P_3 P_4 P_5 = (1/2)(1/2)(1/2)(1/2)(1/2) = 1/32$$

Multiplying fractions or decimals less than 1 or percentages less than 100 results in a dramatic decrease in the outcome of an equation, event, or process. This fact is used by nuclear engineers when they design nuclear power plants, as part of risk management. They purposely (and thoughtfully) create a design that has an extraordinarily low probability for a catastrophic event to occur by precluding such a catastrophic accident unless several low-probability failures have occurred simultaneously or sequentially. By being sure that individual links in the chain are not inherently interrelated or interlinked, the overall probability can be made extremely small, and a catastrophe, nearly impossible.

For example, for just three links with probabilities of failure of 1/1000, 1/10,000, and 1/1,000,000, the overall probability of all three occurring simultaneously or sequentially is:

$$P_{overall} = P_1 P_2 P_3 = (10^{-3})(10^{-4})(10^{-6}) = 10^{-13}$$

That's 1 in 10 trillion! Possible, but extremely unlikely.

Illustrative Example 11-4: Terms Raised to a Power

Bullets shot from a rifle or handgun have a kinetic energy (as they leave the muzzle) of $KE = 1/2\,mv^2$, in which m is the mass of the bullet and v is the velocity of the bullet at the muzzle (i.e., the muzzle velocity).

[4] *Shape factor* is well described, along with the methodology for deriving values, in Michael F. Ashby, *Material Selection in Mechanical Design*, 3rd edition, Elsevier, 2005.

For maximum destructiveness (i.e., penetration and/or stopping power), the KE should be as high as possible. From the equation for KE, it is clear that muzzle velocity is much more important than bullet mass, since the velocity term is squared.

Note for a .308 Winchester hunting rifle (which is comparable to the 7.6×50mm military standard for NATO), the greater effect of muzzle velocity (MV) compared to bullet weight or mass (W_b) on muzzle energy (ME) can be seen from the following data:

	Bullet Mass (grains)	MV (fps)	ME (ft-lbf)
.308 Winchester	165	2685	2641
.308 Winchester	165	**2870**	**3019**
.308 Winchester	180	2620	2743

When the U.S. Army developed a "bunker buster" bomb that could penetrate 20 ft of reinforced concrete in an effort to kill Sadaam Hussein of Iraq, they were forced to seek the densest material for the penetrating tip (i.e., the most massive projectile), as the velocity would be limited to the terminal velocity for the nonpropelled, free-falling bomb.

A final example is a little different, on first glance, but still involves sensitivity analysis.

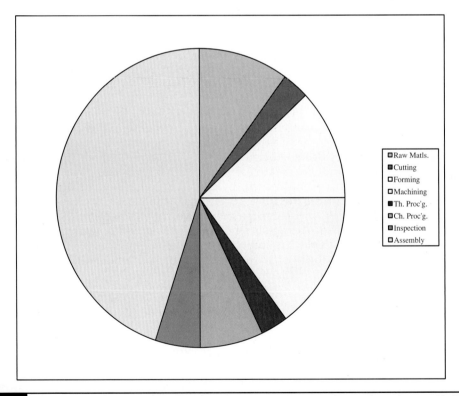

Raw Matls.
Cutting
Forming
Machining
Th. Proc'g.
Ch. Proc'g.
Inspection
Assembly

 Figure 11-1 A pie chart showing the relative contributions to the cost of manufacturing an Al alloy airframe for the major elements of manufacture of an airplane. Cost elements as segments in the pie chart clockwise from 12 o'clock correspond to the order listed in the key from top to bottom.

Illustrative Example 11-5: Another Sensitivity Analysis

Pie charts plot the contribution of each of multiple factors as segments of a circle (i.e., "pieces of a pie") (see Chapter 26, "Graphing and Graphical Methods").

As a young engineer charged with reducing the cost of manufacturing airframes for military airplanes, the author, as a materials engineer working in "advanced materials and processes development" was inclined to seek less costly materials. However, the director of advanced development, who had decided to be a good mentor to create a better development engineer, encouraged the author to "look into the cost elements associated with airframe manufacturing, in the form of a pie chart."

The result of the author's effort is represented by Figure 11-1.

In presenting the pie chart to the director, the author was encouraged to "go for the biggest piece of the pie." He correctly pointed out that if the cost (via manual "touch" labor) associated with assembly (using upset rivets in Al alloy detail parts) could be reduced by one-fifth, it would represent a greater savings than eliminating the cost of materials altogether.

The lesson: Always go for the biggest piece of a pie!

In Summary

It is extremely important for an engineer to perform a *sensitivity analysis* on equations in order to determine which input parameter(s) has(have) the greatest effect(s) on the output of the equation. To do this, one needs to consider the form of the equation, looking for additive or subtractive factors, multiplicative factors, or factors raised to a power greater than 1.

CHAPTER 12

Response Curves and Surfaces

THE DICTIONARY[1] DEFINES *response,* as a noun, as "a reaction, as that of an organism or a mechanism, to a specific stimulus." Engineers earn a living dealing with responses, either seeking a specific needed or desired response for a specific input or avoiding an undesired or intolerable response for some unknown or for some known or possible but uncontrolled input or stimulus. The response (or responses) an engineer seeks to obtain or, in some cases, to avoid, tends to be one of these two generic types:

1. To achieve a high degree of response for a small change to an input

and/or

2. To avoid a dramatic unwanted response for a small or an uncontrolled input or variation to a known input

The former situation is known as *responsiveness*. The latter situation is known as a *stability* or *robustness*. A couple of examples should help.

An electrical engineer may have a need (or desire) to detect and/or monitor a very weak signal (e.g., a microvoltage), or an astronomer may wish to detect an extremely low-intensity level of light from a distant galaxy. In both cases, the weak input signal needs to be greatly amplified using some type of *gain* device (e.g., an electronic amplifier or light-gathering mirror or system of mirrors). A fighter pilot wants to have rapid but precise maneuverability for small intentional inputs to airfoil control surfaces and the engine, as does a race car driver for small intentional inputs to the steering mechanism and the accelerator. On the other hand—and, sometimes, at the same time—both electrical engineers and astronomers want to have minimal interference from superfluous "noise." Likewise, fighter pilots and race car drivers want stability from unexpected interferences, such as wind gusts and bumps.

By proper design, users, such as these, can have both.

To assess the reaction of a material, device, mechanism, structure, process, or system to inputs, as stimuli, engineers use response curves for 2D problems and response surfaces for 3D and higher problems. From a general standpoint, a *response surface* is a surface in $n+1 > 2$ dimensions that represents the reactions to, as variations in, the expected value of output or *response variable* as the values of n input or *explanatory variables* are changed. In most cases, the interest is in finding the combination that gives a global optimum (i.e., maximum or minimum) response.

To find a maximum response, a common interactive procedure is to seek the steepest ascent (or descent) in the response variablefor explanatory variables that are collinear. To find a minimum response, one seeks a point (or points) where the slope is zero, mathematically determined by taking the derivative of the equation for the response curve.

In statistics, *response surface methodology* (RSM) explores the relationships between several explanatory variables and one or more response variables. This is done using a sequence of designed experiments to obtain an optimum response. There are several special experimental designs for efficient evaluation near a supposed or expected optimum with n explanatory variables. A *control composite*

[1] Definitions are from Houghton Mifflin's online dictionary at www.thefreedictionary.com/Houghton.

design consists of observations at the vertices of a hypercube (i.e., a shape with *n* axes, for $n > 3$, each of equal length), together with repeated observations at the origin and observations on each axis at a distance *c* from the origin, where $c = \sqrt{n}$ for a rotatable design. A Box-Behnken design uses fewer observations by replacing the observations at vertices by observations at the midpoints of edges.[2]

One of the most common areas in which engineers are concerned about response surfaces is assessing *process stability.* A manufacturing process cannot be placed into a production environment until it has been proven to be stable. One cannot even begin to consider process capability until one has demonstrated process stability. A process is said to be stable when all of the response parameters used to assess or measure the process have both constant mean values (i.e., means) and constant variances over time, and also have a constant distribution.

A graphical tool (see Chapter 26) used to assess process stability is *scatter plots.* In this method, a large number of independent samples (typically greater than 100) from a process are collected and measured over a sufficiently long period of time to be representative of production. This period can be measured by time (e.g., hours or days) or as the number of parts or quantity of material processed. The measured data are then plotted with the sample order on the *x* axis, and the value being assessed as the *y* axis. The plot should exhibit constant random variation about a constant mean if the process is stable. Sometimes *control limits* are superimposed on a scatter plot to ensure that no production points fall outside the control limits.

Figure 12–1 shows a scatter plot for the waiting time between eruptions of the Old Faithful Geyser in Yellowstone National Park, Wyoming, USA. From it, one can gather useful information from patterns in the scattered data points. Here, it can be seen that longer waiting times result in longer-duration eruptions, likely the result of greater built-up pressure before venting. While an event, not a process, scatter plots still provide useful information on expected behavior in the future from behavior in the past for this famous geyser.

It is interesting to look at how Nature operates with large systems and complex organisms or processes to fully appreciate response to stimuli in mechanisms or processes. What were once believed to be random occurrences in the natural world, such as weather or how branches form on trees, have been found to be anything but random.

Chaos theory is a field of mathematics with broad applications in physics and biology, as well as in economics and philosophy. Chaos theory studies the behavior of dynamical systems that are extremely sensitive to initial conditions—popularized as the "butterfly effect."[3] Small differences in initial conditions (such as those due to rounding errors in numerical computations; see Chapter 4) yield widely diverging outcomes for chaotic systems. These generally make long-term prediction impossible, even though the processes and/or systems involved are determinate—in that their future is fully determined by their initial conditions when no random elements are involved. Being determinate, however, does not make chaotic processes predictable.

Illustrative Example 12–1: Trying to Influence Behavior

Hurricane Katrina formed over the Bahamas on August 23, 2005, and crossed southern Florida as a moderate Category 1 tropical storm, leaving behind some flooding and relatively few deaths. When it reached the warm Gulf of Mexico west of the Florida peninsula, it quickly gathered energy and strength, building to a deadly Category 4 hurricane just before making landfall along the northern Gulf Coast at Alabama, Mississippi,

---

[2]The reader interested in experimental design should look into "factorial experiments" for more information.
[3]The "butterfly effect" suggests that a very minor perturbation to an initial condition, such as to the air surrounding a butterfly when it flaps its wings, has effects well beyond the immediate vicinity of the butterfly—in fact, it affects the entire atmosphere of Earth.

Old Faithful Eruptions

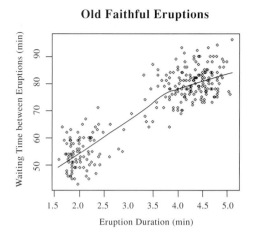

Figure 12-1 A scatter plot of waiting times between eruptions of Old Faithful Geyser in Yellowstone National Park, Wyoming, USA. It can be seen that waiting times seem to cluster around two mean times of 50 to 60 minutes and 75 to 85 minutes, with longer waiting times resulting in longer-duration eruptions, likely from greater built-up pressure before venting. (*Source:* Wikimedia Commons; originally contributed by Maksim on 20 March 2006 and released by this author to the public domain for free use.)

and Louisiana. It struck New Orleans a devastating blow on August 28, and, while weakening to a Category 3 hurricane, still became one of the costliest natural disasters, and one of the five deadliest hurricanes, in the history of the United States. Poor response by too many federal, state, and local officials to name led to a barrage of postdisaster attacks and accusations.

One well-meaning, but scientifically naïve, environmental activist (with an underlying political agenda) suggested the deadly storm could have been "deflected away from New Orleans" had the federal government taken proper action. He proposed that carbon-black soot could have been sprinkled over the building cyclonic storm, thereby "increasing the absorption of heat from the Sun" (above the clouds), thereby "increasing updraft" and "causing the storm to take a different path" so as "to miss New Orleans." The problem is: How does one change the path of a hurricane whose behavior is chaotic and unpredictable in any meaningful and controlled way? It is equally likely that the proposed idea could have caused the eye of the storm to hit New Orleans when it would not have had it not been tampered with!

There are two lessons: First, one cannot meaningfully influence and/or control a chaotic event. Second, to quote—or paraphrase—an old proverb, "A little knowledge [of science] is a dangerous thing."

In many situations, and for many circumstances, an engineer needs to know how a particular mechanism, process, or system will respond. For processes used in manufacturing, the desire for stability indicates that a response curve, as a 2D plot of an outcome variable to an input variable, has a low slope or, perhaps, lies at a relative minimum (i.e., an actual minimum, as determined by the derivative of the equation for the response parameter in terms of the input parameter being zero, or an inflection point, as determined by the second derivative being zero). Examples of such points of stability for some generic response curves are shown in Figure 12–2. Undesirable locations, where change in the response is dramatic for a small change in the input parameter, are also shown.

Two examples of how engineers seek stable regions of a 2D response curve are given in Figure 12–3a and *b*. Because of the normal statistical variation of the peak stress that will lead to a given number of

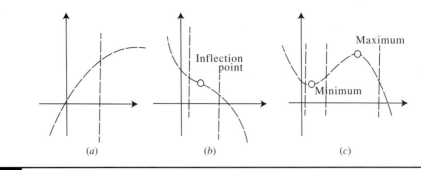

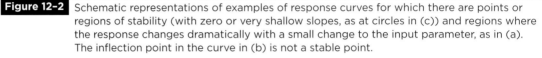

Figure 12-2 Schematic representations of examples of response curves for which there are points or regions of stability (with zero or very shallow slopes, as at circles in (c)) and regions where the response changes dramatically with a small change to the input parameter, as in (a). The inflection point in the curve in (b) is not a stable point.

cycles of expected life during cyclic (fatigue) loading of a structure, along with the statistical variation in the number of cycles of expected life for a particular peak stress, design engineers tend to try to operate below the material's *fatigue limit,* below which stress there should be limitless life. For applications in which materials (and the parts they make up) operate above about one-third of their absolute melting temperature (i.e., a homologous temperature, $T/T_{MP} > 0.33$), a material (and part) undergoes continuous strain in what is known as *creep.* For predictability, design engineers seek to operate in the region of steady-state behavior (i.e., within the secondary creep region), where the strain rate is a minimum and constant.

For situations in which there are multiple input parameters or explanatory variables, *response surfaces* are used rather than response curves. An example for a hypothetical response influenced by two explanatory variables, Factor A and Factor B, is shown in Figure 12–4. For relative stability, one would try to operate in a region defined by *control limits* that would restrict responses to a relatively flat portion of the surface—here, the region "10–12," but possibly extending out into region "8–10." There are many real-life examples in processing, as well as in trials with new pharmaceuticals.

While some response curves and response surfaces are generated by mathematical equations, others are the result of experimentation or sampling. However, curves or surfaces for such cases can be represented by mathematical equations generated by regression analysis. From this standpoint, response

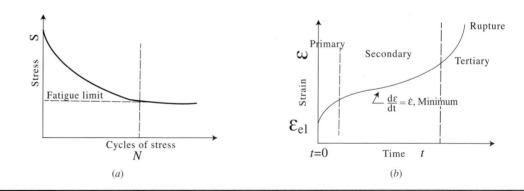

Figure 12-3 (*a*) A typical *S-N* curve used to assess the effect of cyclic loading on the life of a material or structure, i.e., fatigue behavior. Stability is reached when the applied stress is below the *fatigue limit.* (*b*) A typical creep curve of strain versus time for a material or structure operating above about one-third of its absolute melting (i.e., homologous) temperature. Steady-state creep occurs in the "secondary" region, where the strain rate is a minimum and constant.

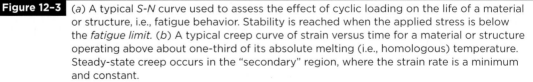

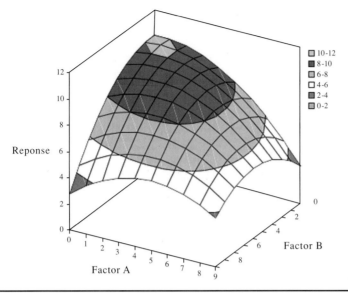

Figure 12-4 An example of a hypothetical response surface for two input or explanatory variables, Factor A and Factor B. Stability is greatest where the slopes in all directions on the surface are lowest. The key represents regions of the response surface from top to bottom. (*Source:* "Response Surface Methodology"—from Google search—by David Harvey is licensed under Creative Commons Attribution Non-Commercial-Share-Alike 3.0 Unported, and is free for use.)

curves and response surfaces can be viewed as either a mathematical problem-solving technique or a graphical technique (see Chapter 26).

A couple of illustrative examples for the problem-solving technique of using response curves and surfaces follow.

Illustrative Example 12-2: Using a Response Curve

As a design engineer, you are faced with selecting a zinc alloy for a small cast-metal part in a mechanical assembly. Your basis for focusing on zinc-based alloys is their low melting point, and associated low mold cost, and good castability. The part is to operate at room temperature (RT) and will be subjected to a maximum stress of 175 MPa (~25,000 lb/in²) for a total time of about 100 hours. The majority of time, the part operates at a fraction of this stress. With melting points in the range of 350 to 420°C (623 to 693 K), zinc alloys at room temperature (25°C or 298 K) are at a homologous temperature (T/T_{MP} absolute) of 0.43—0.48. At this level, they will thus experience creep under sustained stresses.

You find creep curves for three zinc alloys (Figure 12-5) on the Internet (at www.dynacast.com), and you choose ACuZinc5 for two reasons: First, the data show it can operate up to 200 MPa (~29,000 lb/in²) at RT for well more than 100 h. Second, even though alloy ZA8 might survive at 175 MPa for design (versus for the 200 MPa for which data are shown), the creep rate is very high and on a steeply rising portion of the curve. The alloy ACuZinc5, on the other hand, will be operating in the stable steady-state region for the conditions required of your design.[4]

[4]Designers always prefer to use an alloy subjected to creep in its steady-stage regime for both safety and predictability.

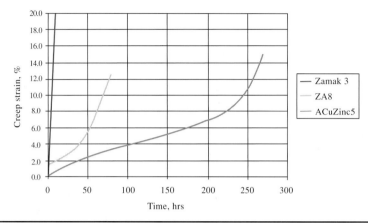

Creep Curves for Zinc Alloys
(Room Temperature & 200 MPa, 29,000 lb/in^2)

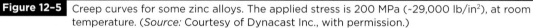

Figure 12-5 Creep curves for some zinc alloys. The applied stress is 200 MPa (~29,000 lb/in^2), at room temperature. (*Source:* Courtesy of Dynacast Inc., with permission.)

Illustrative Example 12-3: Using a Response Surface

You conduct a series of experiments to assess the effects of two input parameters on an output response. You get a response surface that exhibits a "saddle-point" like that shown in Figure 12–6. Even though the region around the saddle-point is relatively flat, you are concerned because deviation from control limits in one direction will lead to rapid decline,

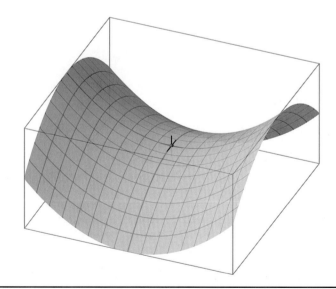

Figure 12-6 A response curve exhibiting a "saddle-point," where a maximum exists in some directions and a minimum in other directions. The equation for the surface shown is $z = x^2 - y^2$. (*Source:* From a Google search, "Response Surface Methodology" by David Harvey is licensed under Creative Commons Attribution Non-Commercial-Share-Alike 3.0 Unported, so is free to use.)

even though deviation in the other (orthogonal) direction suggests stability (as a relative minimum point in that plane).

You correctly decide that a saddle-point on a response surface is not a good place to operate a manufacturing process.

In Summary

How quickly a response changes as input parameters (or explanatory variables) change is expressed by 2D or 3D plots to produce what are known as *response curves* and *response surfaces,* respectively. Whether viewed as a mathematical approach or a graphical approach to problem-solving, response curves and surfaces are useful for many situations in engineering.

CHAPTER 13

Numerical Analysis and Methods

IT HAS BEEN estimated by experts that more than 99.9 percent of real-world problems in engineering and science are so complicated that they cannot be solved by analytical methods for which there are exact, closed-form (analytic) solutions. In other words, the math learned throughout primary and secondary school (e.g., algebra, geometry, trigonometry, and precalculus) and in required math courses in college or engineering school (e.g., calculus, differential equations) is enough only in rare instances. Nevertheless, the math one has learned provides the core truths, and, most important, helps with understanding of the mathematical problems encountered in science and engineering and allows testing of the integrity of what are known as *numerical* (versus analytical) *schemes*. In short, most problems in engineering and science can only be solved numerically using what is known as *numerical analysis* employing *numerical methods*. For this reason, numerical analysis and numerical methods are an extremely important problem-solving technique in engineering and science, as they are in other fields (e.g., economics) as well.

The overall purpose of the field of numerical analysis is to design and assess techniques that give approximate, but quite accurate, solutions to complicated problems. The variety of such problems can be appreciated from the following examples:

- A successful space program requires accurate numerical solution of the set or system of ordinary differential equations for computing the trajectories of spacecraft—as these are affected by Earth, the Sun, the Moon, and, for deeper probes, the other planets and, eventually, stars.
- Modern automobile manufacturers use computer simulations of car crashes to improve safety through better crashworthiness, in which partial differential equations need to be solved numerically.
- Meteorologists attempt to predict weather using advanced numerical methods, including complex equations for chaotic behavior.
- Airlines use sophisticated algorithms to schedule flights, assign aircraft and crews, calculate fuel needs, and decide on airline ticket prices.

As important as numerical analysis and methods are as mathematical approaches to problem-solving in engineering, it is far too extensive and complicated a subject to properly cover in a book like this (i.e., on the spectrum of problem-solving techniques). Fortunately, there are a host of excellent books devoted entirely to numerical analysis and, more usually, to numerical methods (see "Suggested Resources" at the end of this chapter). Conveniently, there are also a number of excellent sources on the Internet, usually as online lecture notes by university faculty who teach the subject (see "Online References").

While the field of numerical analysis predates the invention of modern computers by many centuries,[1] the invention of the computer has had a profound impact on the speed with which these notoriously tedious iterative methods can be computed *and* on the complexity of calculations and problems that

---

[1]Linear interpolation methods date back at least 3000 years, with evidence for much early use (e.g., in finding a precise value for $\sqrt{2}$ using a graphical technique).

can be undertaken. Prior to the advent of modern computers, numerical methods depended on hand calculations involving interpolation.

Areas of study in which numerical analysis is used are divided according to the problem that is to be solved, with major areas being:

■ Computing values of certain functions
■ Interpolation, extrapolation, and regression[2]
■ Solving equations and systems of equations (e.g., systems of linear equations and root-finding for nonlinear equations)
■ Solving eigenvalue and singular-value problems
■ Optimization (or mathematical optimization)
■ Evaluating integrals (or numerical integration)
■ Computing the solution of both ordinary and partial differential equations

Illustrative Example 13–1: Simple Examples of Typical Areas

a. *Interpolation:* An express train moving at a constant speed of 50 mph (80 kph) that passes a station at 3:15 p.m. CST passes a second station 75 mi (120 km) away at 4:45 p.m. CST. A police officer investigating an incident that occurred along the track between the two stations wants to know what time the train would have passed a mile marker 32 mi (51 km) from the first station.

b. *Extrapolation:* If a company's sales have increased an average of 7 percent per year for each year over the past five years, reaching $31 million, the CEO wants to estimate sales in each of the next three years if the growth rate remains constant. (Compounding must be considered, as after one year, sales would be 1.07 × $31 million, but sales after another year has passed would be 1.07 × 1.07 × $31 million, and so on.)

c. *Regression:* In linear regression (or regression analysis), a straight line can be computed that passes as close as possible to every one of the points in a collection of *n* points. (This is routinely done to fit curves to measured data points using the method of least squares.)

d. *Optimization:* If a company sells a product at $100, it can sell 1000 units per month. Market research shows that for every $5 increase in price above $100, 50 fewer units per month would be sold. The question is: What price should the company charge for the product if it wishes to maximize its gross sales per month?

e. *Differential equations:* If you set up 20 fans to blow air from one end of a large room to the other end and drop a feather into the wind, the feather will follow the air currents, which would likely be very complex. One approximation would be to measure the speed of the feather every second, advance the blown feather as if it was moving in a straight line at that speed for 1 s, and then measure the speed of the feather (i.e., the wind speed) anew. This is known as the Euler method for solving an ordinary differential equation.

Each of these preceding examples is far easier to solve numerically than analytically.

Illustrative Example 13–2: An Analytical versus a Numerical Solution

The *analytic solution* for a general quadratic equation given by

$$ax^2 + bx + c = 0$$

---

[2]Interpolation and extrapolation are discussed in Chapter 5. Regression is similar, but takes into account whether the data is imprecise, the least-squares method being a particularly popular approach.

is

$$x = -b/2a + (b^2 - 4ac)^{1/2}/2a$$

While the solution works for any set of values a, b, and c, real solutions exist only when $b^2 - 4ac \geq 0$. The properties of the solution are clear.

Using a numerical approach (i.e., seeking a numerical solution), one can only deal with a given set of a, b, and c at a time, the solution will be approximate, and an estimate of error is needed. As for all mathematical problems, it is useful at the start to try to visualize the problem before trying to solve it. This can be aided by considering the form of the equation. Here, the equation $f(x) = ax^2 + bx + c$ is a parabola for which real solutions to the specific equation $ax^2 + bx + c = 0$ are the intersections of the parabola with the zero line or x axis at $f(x) = 0$ (Figure 13–1).

Not all quadratic equations can be factored to give real roots or solutions. Those quadratics for which there are no real roots have curves that do not intercept the x axis.

Using a numerical approach, if X is a solution to the equation, and x_1 and x_2 are points immediately to the left and to the right of X, $f(x_1)$ and $f(x_2)$ will have opposite signs and a solution falls within the interval (x_1, x_2). The goal of a numerical scheme is to systematically refine or narrow this interval. A "bisection scheme" would approach the solution by:

1. Picking two points x_1 and x_2 such that $f(x_1)$ and $f(x_2)$ have opposite signs—often needing to be done manually.
2. Bisecting the interval (x_1, x_2) into (x_1, x_m) and (x_m, x_2), where x_m is the midpoint of the original interval, and keeping the half-interval intervals for which $f(x)$ have opposite signs at the two endpoints.
3. Repeating Step 2 until the refined interval is short enough for the needed accuracy. The midpoint of this last interval is the numerical solution and the interval is the *error bar*.

A computer helps with the repetitive computation associated with repetitive bisections.

The central idea behind many methods used in numerical analysis is the *Taylor series expansion* of a function about a point. In mathematics, a *Taylor series* is a representation of a function as a sum of infinite terms that are calculated from the values of the function's derivatives at a single point. In common practice, a function is approximated by using a finite number of terms in its Taylor series (Figure 13–2).[3] The result is called a *Taylor polynomial*. Taylor's theorem gives quantitative estimates of the error in such approximations. Not every function can be so approximated, as there are conditions that must apply to make functions suitable to the method.

Two numerical methods meriting special mention are: (1) the finite element method and (2) the finite difference method. The *finite element method* (FEM)[4] is a numerical technique for finding approximate solutions of partial differential equations, as well as of integral equations. The approach to a solution requires either eliminating the differential equations completely by imposing a steady-state condition (see Chapter 15) or approximating the partial differential equation with a set of ordinary differential equations. The latter can be numerically integrated using any of a number of standard techniques (e.g., Euler's method). Interested readers are encouraged to seek specific references on the topic.

FEM is used extensively in aeronautical, biomechanical, mechanical, and electrical engineering, particularly in design situations in which thermal, fluid, structural, and/or electromagnetic working environments are involved.

---

[3] For a function containing a single variable x, i.e., $f(x)$, the Taylor series expansion may be represented as: $f(x + \delta x) = f(x) + \delta x\, f'(x) + \delta x^2/2\, f'(x) + \delta x^3/6\, f''(x) + \ldots$.
[4] The practical application of FEM is generally known as *finite element analysis* (FEA).

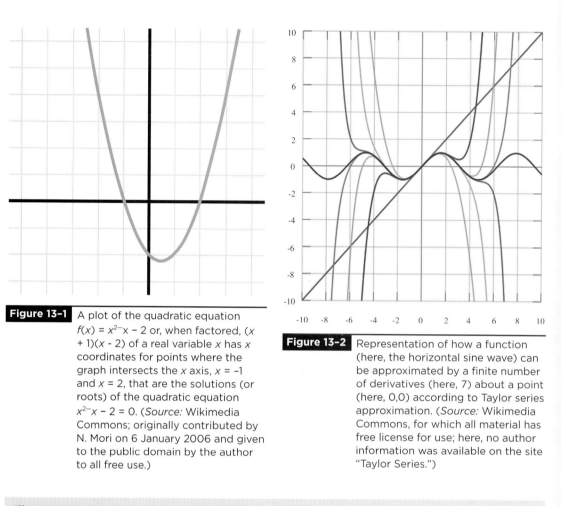

Figure 13-1 A plot of the quadratic equation $f(x) = x^2 - x - 2$ or, when factored, $(x + 1)(x - 2)$ of a real variable x has x coordinates for points where the graph intersects the x axis, $x = -1$ and $x = 2$, that are the solutions (or roots) of the quadratic equation $x^2 - x - 2 = 0$. (*Source:* Wikimedia Commons; originally contributed by N. Mori on 6 January 2006 and given to the public domain by the author to all free use.)

Figure 13-2 Representation of how a function (here, the horizontal sine wave) can be approximated by a finite number of derivatives (here, 7) about a point (here, 0,0) according to Taylor series approximation. (*Source:* Wikimedia Commons, for which all material has free license for use; here, no author information was available on the site "Taylor Series.")

Illustrative Example 13-3: Using FEM

An excellent tutorial example to show the use of the finite element method (FEM) can be found via a Yahoo search under "FEM example with traffic lights on beam." The example created by Eron Flory '97 under the advisement of Professor Joseph J. Rencis deals with the familiar situation of two traffic lights mounted to a horizontal beam that is cantilevered from a vertical support pole (Figure 13-3).

The tutorial goes through the details of calculating the vertical displacement and rotation of the horizontal beam at the traffic light mounts, the support reactions, and the internal forces at the light mounts near the center of the beam.

Check it out for yourself.

In mathematics, *finite difference methods* (FDM) are numerical methods for approximating the solutions to differential equations using equations of finite differences to approximate derivatives. The method generally involves the use of Taylor polynomials. To use FDM to approximate a solution to a problem, it is first necessary to divide the domain of the problem into a uniform grid (i.e., to discretize the problem's domain). The finite difference method produces sets of discrete numerical approximations to the derivative in what is known as *time stepping* (Figure 13–4).

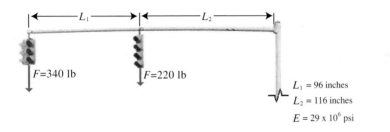

F=340 lb F=220 lb

L_1 = 96 inches
L_2 = 116 inches
E = 29 x 10^6 psi

Figure 13-3 This example problem, for which numerical solution works best, was created by Eron Flory '97 under the advisement of Professor Joseph J. Rencis (formerly of the Department of Mechanical Engineering at the University of Arkansas, and now Dean of Engineering, Clay W. Hixon Chair for Engineering Leadership and Professor of Mechanical Engineering at Tennessee Technological University). Permission for use was kindly granted by both Eron Flory and Professor Rencis.

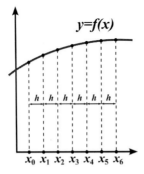

$y=f(x)$

h h h h h h

x_0 x_1 x_2 x_3 x_4 x_5 x_6

Figure 13-4 Time-stepping in the finite difference method. Here, six steps are used to approximate the function $y = f(x)$. (*Source:* Wikimedia Commons; originally contributed by Mintz 1 on 23 September 2006, later modified by Nicoguardo on 21 February 2011, used free, as file was released by the original author to the public domain.)

The error between the approximate and the exact analytical solution arises from (1) the loss of precision from rounding decimal quantities and (2) the error from discretizing the problem into a finite (versus infinite) number of steps.

One of the most common areas for applying the FDM is for solving the generalized heat equation.[5]

Illustrative Example 13-4: Using FDM

An example of the use of the finite difference method for a heat flow problem is given in "A Limited Tutorial on Using Finite Difference in Soil Physics Problems," by Donald L. Baker, on www.aquarien.com/findif/Findifa4.html.

Check it out for yourself.

---

[5]The general equation of heat flow is given by: $\partial u/\partial t - \alpha \, (\partial^2 u/\partial x^2 + \partial^2 u/\partial y^2 + \partial^2 u/\partial z^2) = 0$ for heat flow (as a function u) in three spatial dimensions (x,y,z) with time (t).

In Summary

The overwhelming majority of problems encountered in engineering are too complex to solve analytically. These require numerical solution via *numerical analysis* and using *numerical methods.* Solutions are approximate, but the error compared to a precise analytical solution (if one were obtainable) can be reduced to whatever level is desired provided one can put up with tedious iterative calculations. Modern computers apply the brute force needed, and do so in greatly reduced time frames. Two of the most common numerical methods are the *finite element method* (FEM) and the *finite difference method* (FDM).

Suggested Resources

Books

Huebner, Kenneth H., Donald L. Dewhurst, Douglas E. Smith, and Ted G. Byrom, *The Finite Element Method for Engineers,* 4th edition, John Wiley & Sons, Hoboken, NJ, 2001.

LeVeque, Randall J., *Finite Difference Method for Ordinary and Partial Differential Equations,* Society for Industrial and Applied Mathematics, Philadelphia, PA, 2007.

Logan, Daryl L., *A First Course in Finite Element Methods,* 5th edition, Cengage Learning, Independence, KY, 2011.

Ozisik, M. Necati, *Finite Difference Method in Heat Transfer,* CRC Press, New York, 1994.

Online References

Gillow, Kathryn, A Finite Element Method Tutorial, at http://www.cs.ox.ac.uk/people/Kathryn.gillow/femtutorial.pdf.

Zhilin Li, Finite Difference Method Basics, at http://www.ncsu.edu/~zhilin/TEACHING/MA402/notes.pdf.

CHAPTER 14

Dimensionless Quantities or Parameters

IN DIMENSIONAL ANALYSIS (see Chapter 3), a *dimensionless quantity* is a quantity without any associated physical dimensions. Thus, it is a pure number with a dimension of 1 (i.e., it is a quantity of dimension one). *Dimensionless quantities* or *parameters* are widely used in mathematics, physics, and engineering. They are also used in everyday life when we count (e.g., 1 [apple], 2 [apples], 3 [apples], etc., the particular item being counted being taken for granted, as opposed to identified each time).

Dimensionless quantities are often defined by the product and/or ratio of quantities that have dimensions (i.e., that are not dimensionless). For such products or ratios to be dimensionless, the dimensions of the quantities involved in the products or ratios must cancel out. A good example is engineering strain as a dimensionless measure of deformation. *Engineering strain* is defined as a change in length ΔL divided by an initial length L_o (i.e., $\Delta L/L_o$). Since both are lengths, with length dimensions, engineering strain is dimensionless; e.g., mm/mm, m/m, or in/in, which are all equivalent as ratios. Surely the best-known dimensionless quantity is π (also known as Archimedes's ratio), the mathematical constant that derives from the ratio of the circumference of a circle to its diameter.

Dimensionless quantities have a couple of important properties. First, even though they have no physical dimension themselves, they can still have dimensionless units, with the same units in both the numerator and the denominator. Examples, besides engineering strain and π, include concentrations (e.g., kg/kg or mol/mol) and angles of measure (in degrees or radians) as the ratio of portions of a circle to a full circle (with 360 degrees or 2π radians). Second, a dimensionless ratio of two quantities with the same units has the same value regardless of the units used to calculate the quantities. As an example, when an object A exerts a force of magnitude F on another object B, and that second object B exerts a force of magnitude f on the first object A, the ratio F/f (or f/F) will be 1 regardless of the actual units used to express the magnitude of the forces F and f. This is a key property of dimensionless ratios and arises from the assumption (see Chapter 16) that the laws of physics are independent of the system of units used to express them, and is closely related to the Buckingham π theorem. The *Buckingham theorem* states that any physical law can be expressed as an equation that is always true (i.e., is an *identity*) involving only dimensionless combinations (as products or ratios) of the variables linked to the law.

A consequence of the Buckingham π theorem used in dimensionless analysis is that the functional dependence between n variables can be reduced by the number k of independent quantities to give a set $p = n - k$ independent dimensionless parameters. This allows different systems that show the same physical description by dimensionless parameters to be equivalent. Following is an example.

Illustrative Example 14-1: Using Buckingham's π Theorem

If the blade or paddle in a mixing apparatus has a given shape, the power needed to operate the mixer is a function of (1) the density and (2) viscosity of the fluid to be mixed,

(3) the size of the mixer blade (as its diameter), and (4) and (5) the speed of the mixer blade as distance divided by time, in other words, five variables ($n = 5$). These five variables are made up of three dimensions ($k = 3$): length L (m), time t (s), and mass m (kg).

According to the π theorem, these $n = 5$ variables can be reduced by the $k = 3$ dimensions to create $p = n - k = 5 - 3 = 2$ independent dimensionless numbers, which are:

- Reynolds number (Re), a dimensionless number that describes the fluid flow
- Power number or Newton number Np, a dimensionless number that describes the mixer and also involves the density of the fluid

There are many dimensionless quantities of importance in engineering and science listed in wikipedia.com under "Dimensionless Quantities." Table 14–1 gives the author's short list of some of the most commonly used dimensionless quantities, along with their respective field of application.

Illustrative Example 14–2: Using the Reynolds Number (Re)

The Reynolds number (Re) expresses the ratio (see Chapters 8 and 9) between total momentum transfer and molecular momentum transfer in fluids. Alternatively, it is the ratio of inertia forces to viscous forces.[1] It is used to characterize flow regimes in fluids, most often to distinguish between *laminar flow* and *turbulent flow.* Laminar flow is characterized by smooth, constant flow motion, while turbulent flow is characterized by chaotic eddies, vortices, and other flow instabilities (Figure 14–1). Laminar flow tends to occur below Re values of around 2000, while turbulent flow tends to occur above Re values of 4000. In between, flow is described as transitional flow.

The most common use of the Reynolds number is to determine flow equivalence or dynamic similitude (see Chapter 8).

The formula for the Reynolds number is:

$$Re = \rho v D / \mu$$

in which ρ is the density of the fluid (in either kg/m³ or slugs/ft³), v is the velocity (in m/s or ft/s), D is the characteristic linear dimension (often symbolized by D_H for hydraulic distance, e.g., the diameter of a round pipe, the chord length of an airplane's wing; in m or ft), and μ is the dynamic viscosity (in kg/m-s or slugs/ft-s).[2] Therefore, Re is dimensionless, as units of terms in the numerator and denominator cancel out, thus:

$$(\text{kg/m}^3)(\text{m/s})(\text{m})/(\text{kg/m-s}) = 1$$
$$(\text{slugs/ft}^3)(\text{ft/s})(\text{ft})/\text{slugs/ft-s}) = 1$$

For dynamic similitude (similarity) in three different fluids—water, molasses, and SAE 30 motor oil at 20°C—the value of Re would have to be the same for each during flow through a pipe of hydraulic diameter $D_H = 0.1$ m (i.e., 10 cm), say; that is, $Re_{water} = Re_{molasses} = Re_{oil}$. Given values of density ρ and viscosity μ for these fluids as follows:

	Density ρ (kg/m3)	Viscosity μ (kg/m-s)
Water	1000	998.29
Molasses	1400	11,166
SAE 30 motor oil	890	72,583

[1] Another way to think about what happens when a fluid flows is how much it tries to stay together and move as a unit (i.e., exhibit laminar flow) versus how much it tries to come apart (i.e., exhibit turbulent flow).

[2] The *slug* is the unit of mass in the English (or Imperial) system of units. It is the mass that accelerates by 1 ft/s² when a force of 1 lbf is exerted on it. Thus, it has a mass of 32.17405 lbm or 14.5939 kg.

TABLE 14-1 Some Commonly Used Dimensionless Quantities in Engineering (selected from a comprehensive list on Wikipedia.com)

Quantity	Symbol	Field of Application
Archimedes number	Ar	Motion of fluids due to density differences
Bagnold number	Ba	Flow of bulk solids (e.g., sand and grain)
Bejan number	Be	Dimensionless pressure drop along a channel
Biot number	Bi	Surface vs. volume conductivity of solids
Coefficient of static friction	μ_s	Friction of solid bodies at rest
Coefficient of kinetic friction	μ_k	Friction of solid bodies in transitional motion
Drag coefficient	C_d	Flow resistance
Fourier number	Fo	Heat transfer
Lewis number	Le	Ratio of mass diffusivity to thermal diffusivity
Lift coefficient	C_L	Lift available from airfoil for given attack angle
Mach number	M	Ratio of speed to speed of sound (gas dynamics)
Nusselt number	Nu	Heat transfer with forced convection
Poisson's ratio	v	Elastic lateral to axial dimensional change
Power number	N_p	Power consumption by agitators
Refractive index	n	Speed of light in a material vs. in vacuum
Reynolds number	Re	Ratio of fluid inertial and viscous forces
Stanton number	St	Heat transfer in forced convection
Stefan number	Ste	Heat transfer during phase change

For flow to stop being laminar in each, Re would have to exceed 2000. For these three fluids, the corresponding flow velocities v (m/s) from $\mu Re/\rho D_H$ to begin to transition from laminar flow would be[3]:

$$v_{water} = (998.29 \text{ kg/m-s})(2000)/(1000 \text{ kg/m}^3)(0.1 \text{ m}) = 19.96, \sim 20 \text{ m/s}$$
$$v_{molasses} = (11,166)(2000)/(1400)(0.1) = 159.5 \text{ m/s}, \sim 160 \text{ m/s}$$
$$v_{oil} = (72,583)(2000)/(890)(0.1) = 1,631 \text{ m/s}, \sim 1,600 \text{ m/s}$$

[3] Units, being the same, are omitted for the molasses and the SAE 30 motor oil.

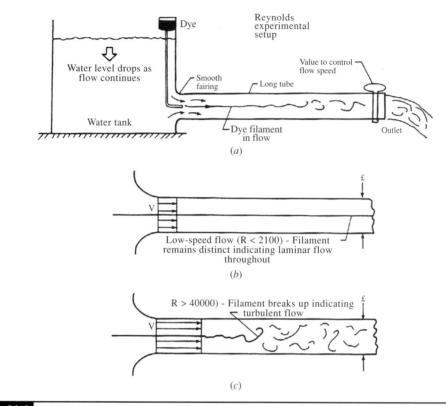

Figure 14-1 Schematic illustration of (*a*) Reynolds' experimental setup for studying fluid flow, and (*b*) the transition from laminar flow at low flow speeds (here, Re < 2100) to (*c*) turbulent flow at high flow speeds (here, Re : 40,000). (*Source:* NASA U.S. Centennial of Flight Commission, under a search for images of "Reynolds number"; used without permission as part of the public domain.)

Before closing, it is important to mention *dimensionless physical constants* that occur in physics as universal constants. Like dimensionless quantities or dimensionless parameters, dimensionless physical constants are independent of the units used. However, they are restricted in their use to quantum physics and cosmology. There are currently 26 *known* fundamental dimensionless physical constants, but there could be more. The interested reader is encouraged to look up "dimensionless physical quantities" on the Internet.

In his book, *Just Six Numbers,* Martin John Rees, a British cosmologist and astrophysicist, ponders six dimensionless constants whose values he believes are truly fundamental to modern physics and the structure of the universe. These six are:

- $N \approx 10^{36}$, the ratio of the fine structure constant (i.e., the dimensionless coupling constant for electromagnetism) to the gravitational coupling constant, the latter of which is defined using protons, as governs the relative importance of gravity and electrostatic attraction/repulsion in explaining the properties of baryonic matter.
- $\varepsilon \approx 0.007$, the fraction of the mass of four protons that is released as energy when fused into a helium nucleus; ε governs the energy output of stars (including our Sun), and is determined by the coupling constant for the strong force.
- $\Omega \approx 0.3$, the ratio of the actual density of the universe to the critical (minimum) density required for the universe to eventually collapse under its own gravity; it will determine the ultimate fate of the

universe. If >1, the universe will collapse in a "Big Crunch." If <1, the universe will expand until maximum entropy is reached in a "Big Freeze."

- $\lambda \approx 0.7$, the ratio of the energy density of the universe, due to the cosmological constant, to the critical density of the universe.
- $Q \approx 10^{-5}$, the energy required to break up and disperse an instance of the largest known structures in the universe, namely a galactic cluster or supercluster, expressed as a fraction of the energy equivalent to the rest mass m of that structure (i.e., mc^2).
- $D \approx 3$, the number of macroscopic spatial dimensions.

In Summary

Dimensionless parameters or *dimensionless quantities* allow things to be compared independent of the physical system of units used to express the quantities involved in the dimensionless quantity or parameter as a product or ratio. In short, they normalize things.

Suggested Resources

Becker, Henry A. Becker, *Dimensionless Parameters: Theory and Methodology*, John Wiley & Sons, New York, 1976.

Rees, Martin John, *Just Six Numbers*, Basic Books, New York, 2001.

CHAPTER 15

Assumptions, Constraints, Conditions, and Cases

REMEMBER SITTING IN the classroom listening to the physics professor's lectures on falling bodies, trajectories of projectiles, conservation of momentum, and seemingly endless other topics? Do you also remember how careful he or she was to set the context for the new phenomenon by stating the *assumptions* that underlay each equation, the *constraints* on the equation's applicability, the *conditions* that applied to examples he or she worked out in front of the class, and the *cases* that were often presented to come closer and closer to actual behavior as more realistic but complicating details were added? You should, because without realizing it at the time, you were being taught to always know (or be aware of) what underlies, influences, limits, and, ideally, benchmarks approaches to and solutions for problems in science and engineering.[1]

Let's look at these important elements of problem-solving one by one.

The dictionary[2] defines the noun *assumption,* as it applies here, as "a fact or statement, as a proposition, axiom, postulate, or notion, taken for granted." Synonyms related to *assumption* help, with important ones being: *given, premise, presumption,* and *supposition.* It is extraordinarily rare that an equation for some physical phenomenon is not based on some underlying assumptions. A few examples are:

- The equation for the horizontal distance d a projectile will travel on Earth (with gravitational acceleration g)—$d = [v_o^2 \sin(2\theta)]/g$—assumes the absence of (i.e., freedom from) drag from the air, which is almost always present to some degree of drag depending on the density of the air as affected by temperature and humidity. It also assumes that the ground over which the projectile is flying is flat. (Later, the initial conditions, that the projectile start at a height of zero with an initial velocity of v_o and an initial angle of θ, will become clear.)
- Momentum (as the product of velocity v and mass m) is conserved so that $m_1 v_1 = m_2 v_2$ assuming collisions are perfectly elastic (between perfectly rigid bodies) with no losses due to deformation. In fact, baseballs (m_1 moving at v_1) are deformed by bats (m_2 moving at v_2) that deform (by deflecting under bending).

The lesson: Always know (or find out) and state (as part of the setup for solving a problem) the underlying assumptions.

The noun *constraint* is defined as "the state of being restricted or confined within prescribed bounds." As opposed to *assumptions,* which limit the applicability of an equation (for example) as a basis for or foundation to the solution from within, *constraints* limit the range of applicability for the solution from outside. As examples:

---

[1] In point of fact, physics teachers do more to instill good problem-solving techniques than anyone in college. They teach units, conversions, significant figures, assumptions, and on and on. Of course, perhaps most important of all, they teach one to always consider the physics that underlies a phenomenon or process. Engineering faculty should only do as well, as, after all, physics underlies everything!

[2] All definitions are from Houghton Mifflin's *American Heritage Dictionary* online at www.thefreedictionary.com.

- The simplest statement of the equation of heat flow Q—$Q = k \, (dT/dx)$—is constrained in its validity and utility to situations in which heat flows in only one direction x (i.e., is one-dimensional), which could be true, and, more subtly, when the thermal conductivity k for the material in which heat is flowing is a constant, independent of the temperature, which is never true. Ideally, for pure 1D heat flow the conductor must have length x but no cross-sectional area (i.e., y and $z = 0$).
- Bragg's law for diffraction of x rays (for example) of wavelength λ striking a periodic array (i.e., in a crystal) of atoms (i.e., scattering centers) on planes separated by a distance d at an incident angle θ—$n\lambda = 2d \sin\theta$, where n is the order of diffraction, often $= 1$—is constrained in its validity by the crystal being thick (so waves scattered by one plane of atoms are constructively or destructively interfered with by waves scattered from a parallel plane, for all planes of the same orientation) *and* by the atoms in the crystal having a simple cubic arrangement (with a single atom of the same species at each corner of a cubic unit cell),[3] and all atoms being the same (i.e., so they have the same scattering effect).

Conditions, as a noun, are defined as "existing circumstances." Conditions set the context for a physical behavior, a property, or a process, for example. As they apply to problem-solving in science and engineering, conditions come in two types: (1) initial conditions and (2) boundary conditions. *Initial conditions* establish how a phenomenon or process begins. *Boundary conditions* fix the problem and its solution to a domain, region, or range.[4] For problems that are to be solved by a mathematical approach or technique, one or the other or both types of conditions need to be defined. In fact, for problems that require the use of differential equations, solutions cannot be found unless the operative boundary conditions are defined and stated.

As an example of each type of condition, there are the following:

- Once again, the horizontal distance d a projectile travels (i.e., its range), assuming no drag, is given by $d = [v_o^2 \sin (2\theta)/g]$, in which v_o and θ are the initial conditions for (muzzle) velocity and firing (or elevation) angle. The effect of initial conditions on such a trajectory is shown in Figure 15–1.
- See Illustrative Example 15–1.

Illustrative Example 15–1: Boundary Conditions in Problem-Solving

The diffusion of one atomic species (as a solute) in another (as the solvent) in the solid state is described by either of two equations known as Fick's first law and Fick's second law. The former applies to conditions of steady state, in which the source of the diffusing species is so large that it can be considered semi-infinite, such that the concentration of the diffusing species does not change (i.e., remains constant), and, furthermore, the sink into which diffusion is taking place (i.e., the solvent) is also so large that it can be considered semi-infinite, since its overall concentration of solute does not change (i.e., remains constant). This results in a constant gradient of composition C over distance x, giving Fick's first law as:

$$J = -D \, (\delta C/\delta x)$$

in which J is the flux (in kg/m^2-s or atoms/m^2-s), D is the diffusion constant or diffusion coefficient (in m^2/s) and represents the ease with which the solute can move/diffuse. Solution of Fick's first law is straightforward.

---

[3] For other arrays, such as face-centered cubic (FCC) or body-centered cubic (BCC), there are additional constraints on what planes with Miller indices (hkl) cause diffraction peaks through constructive interference; i.e., for FCC, h, k, and l must all be even or all be odd, while for BCC, $h + k + l$ must be even.

[4] *Boundary conditions* may be or may include *initial conditions.*

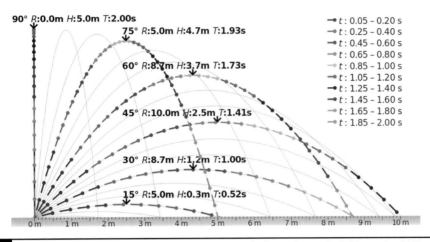

90° R:0.0m H:5.0m T:2.00s
75° R:5.0m H:4.7m T:1.93s
60° R:8.7m H:3.7m T:1.73s
45° R:10.0m H:2.5m T:1.41s
30° R:8.7m H:1.2m T:1.00s
15° R:5.0m H:0.3m T:0.52s

→ t: 0.05 – 0.20 s
→ t: 0.25 – 0.40 s
→ t: 0.45 – 0.60 s
→ t: 0.65 – 0.80 s
→ t: 0.85 – 1.00 s
→ t: 1.05 – 1.20 s
→ t: 1.25 – 1.40 s
→ t: 1.45 – 1.60 s
→ t: 1.65 – 1.80 s
→ t: 1.85 – 2.00 s

0 m 1 m 2 m 3 m 4 m 5 m 6 m 7 m 8 m 9 m 10 m

Figure 15-1 The horizontal distances (i.e., range = R) traveled, maximum height reached (H), and time of flight (T) for equal mass projectiles launched at the same initial velocity (10 m/s) and various elevation angles (in degrees). The time from launch is given by t. The projectile moves through a vacuum under a gravity field of 10/m/s^2, as boundary conditions. (*Source:* Wikimedia Commons; free license; originally contributed by Cmglee on 22 May 2010, used freely under Creative Commons Attribution 3. Share Alike.)

Fick's second law involves a time rate of change in the composition at each and every point x in the sink and a changing concentration gradient for the diffusing species. Thus, it addresses non–steady state and is expressed as:

$$\delta C/\delta x = D\,(\delta^2 C/\delta x^2)$$

This is an ordinary differential equation and, as such, requires *boundary conditions* to be defined in order to arrive at a solution, as there are different solutions for different boundary conditions. A very important solution under non–steady state is a semi-infinite solid sink in which the surface concentration of the solute is held constant. The starting assumptions are thus:

1. Before diffusion begins, any solute atoms in the solvent are uniformly distributed with a concentration of C_o.
2. The value of x at the surface of the solid (solvent) sink is 0 and increases with distance into the solid.
3. Time is measured from 0 at the instant before diffusion begins and increases thereafter.

These assumptions are expressed as boundary conditions as:

For $t = 0$, $C = C_o$ at $0 \le x \le \infty$
For $t > 0$, $C = C_s$ (a constant value at the surface) at $x = 0$
And $C = C_o$ at $x = \infty$

For these boundary conditions, the solution to the differential equation in Fick's second law, to give the concentration at any point x, is:

$$(C_x - C_o)/(C_s - C_o) = 1 - \text{erf}\,(x/2\sqrt{Dt})$$

in which erf $(x/2\sqrt{DT})$ is the gaussian error function for which values can be found in mathematical handbooks.

TABLE 15-1   Cases and Boundary Conditions for Solidification of Alloys

Case 1: Equilibrium Maintained at All Times
    a. $\Delta G_{L-S} = 0$.
    b. No composition gradient exists anywhere; $dc/dx = 0$ in both liquid and solid states.

Case 2: Complete Mixing in the Liquid
    a. Complete mechanical mixing takes place in the liquid, such that no composition gradient ever exists in the liquid; i.e., $dc/dx = 0$ in the liquid.
    b. No diffusion occurs in the solid, so $dc/dt = 0$ in the solid.

Case 3: No Mixing in the Liquid
    a. No mechanical mixing takes place in the liquid.
    b. Changes in composition in the liquid occur by diffusion only.
    c. No diffusion occurs in the solid.

Case 4: Splat Cooling
    a. No mechanical mixing takes place in the liquid.
    b. No diffusion occurs in the liquid, so $dc/dt = 0$ in the liquid.
    c. Liquid is supercooled to a glassy (amorphous) state, rather than to a crystalline state.

Some especially important and common conditions, as bounding conditions, in engineering are:

1. *Equilibrium*[5] versus *nonequilibrium conditions,* as only the former, as completely free of changes, can be subjected to precise analytical description and calculation. For the latter, one needs to attempt to deduce the likely effect on behavior under conditions of equilibrium based on how much the real system (i.e., conditions) differs from equilibrium. The greater the difference, the more unlike the nonequilibrium state will be from the equilibrium state.

2. *Steady state* versus *non–steady state* versus *transient conditions.* Under steady state, a dynamic system comes into balance between inputs and losses (e.g., heat flow from an arc-welding torch moving at a constant speed in an infinite heat sink reaches steady state when the input of new heat is exactly balanced by the loss of heat to produce a fixed temperature distribution around the torch as a function of location for the same increment of time after the torch moves past a given point). Under non-steady state, a dynamic system operates with changing (as opposed to constant) driving force (often as a gradient of temperature, pressure, composition, or electrical potential, etc.). Transients arise from changes to the system, for example, near edges of noninfinite (i.e., finite) masses, during start-up or shutdown of a process, during faults, and so on.

The dictionary defines *cases,* as a plural noun, as "a set of circumstances, state of affairs, or a situation." This sounds similar to conditions and, in fact, involves conditions (i.e., either initial conditions or boundary conditions or both). For engineering, however, as for medicine, law, and business, cases are used to provide useful examples for a problem situation that relates closely to the set of circumstances that seem to apply to the engineer's (or doctor's, lawyer's, or businessperson's) situation. Cases are generally used as a reference point or benchmark.

A good example of the use of cases in engineering appears in theoretical treatments of solidification (i.e., the transformation of a molten material, such as an alloy, into a crystalline solid, as occurs during casting or welding).

---

[5] *Equilibrium* is defined as "a condition in which all acting influences are canceled by others, resulting in a stable, balanced, or unchanging system."

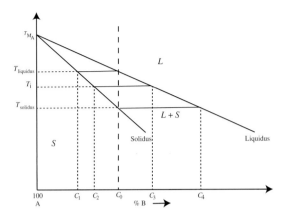

Figure 15-2 Case 1 (Equilibrium Solidification), showing how solute redistributes between the liquid *L* and the solid *S* that is being formed at each instant of time and cooling temperature to form a solid of uniform composition at each instant with no buildup of any solute profile in the liquid ahead of the advancing solid-liquid interface. (*Source:* After Robert W. Messler, Jr., *Principles of Welding,* Wiley-VCH, 1999, Figure 13.16, p. 405; used with permission.)

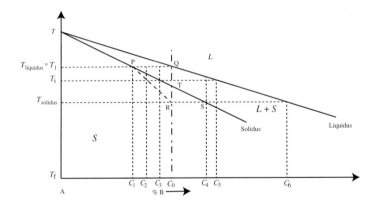

Figure 15-3 Case 2 (Nonequilibrium Solidification/Complete Mixing in the Liquid *L* and No Diffusion in the Solid *S*) showing how the composition of the resulting solid varies from C_1 and the start of solidification at $T_{liquidus}$ to C_4 at a depressed solidus temperature T_f to result in an average composition of C_o in the "cored" structure. (*Source:* After Robert W. Messler, Jr., *Principles of Welding,* Wiley-VCH, 1999, Figure 13.17, p. 408; used with permission.)

Illustrative Example 15-2: Cases in Alloy Solidification

Solidification of crystalline solids (such as pure metals) involves the arrangement (actually, rearrangement) of the atoms of the metal from a state of essential randomness[6] to a state wherein each and every atom occupies a specific point in space to create a structure that repeats along three coordinate axes. In the case of alloys, which are intimate mixtures of

---

[6]In fact, atoms or molecules are completely randomly (but uniformly) distributed in the gaseous state but exhibit some order over short distances (few atom distances) in a liquid (i.e., liquids exhibit short-range order). Crystalline solids exhibit order over long distances (i.e., long-range order).

two or more metallic species, as solidification takes place, all of the atoms in the molten liquid must rearrange into a regular array with long-range spatial order in the crystalline solid and, in addition, have the proper concentrations (or distributions) of each atomic species in the alloy. In all instances, atoms move by the process of diffusion—initially in the high-temperature liquid and, with cooling and solidification, within the ever-cooling new solid.

Treatments of the theory of solidification of alloys (for simplicity, two-component, binary alloys of solute B dissolved in solvent A) have precise models for three cases: one case under conditions of equilibrium (Case 1) and two cases (Case 2 and Case 3) under nonequilibrium that tend to bound reality. A fourth case (Case 4) has been developed for extremely rapid, highly nonequilibrium *splat cooling*. The (boundary) conditions for these four cases are shown in Table 15–1.

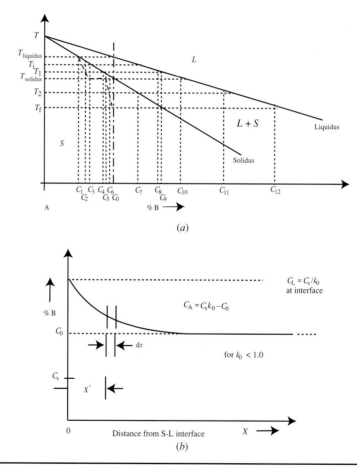

(a)

(b)

Figure 15-4 Case 3 (Nonequilibrium Solidification/No Mixing in the Liquid *L*/No Diffusion in the Solid *S*) showing similar behavior to Case 2, with the creation of a "cored" structure and solidification that is not complete until the average composition of all solid formed reaches C_o (a), but with the additional formation of a solute concentration profile ahead of the advancing solid-liquid interface (b). As a result of the concentration profile in the liquid, constitutional supercooling leads to different growth modes in the solid (*not shown*). (*Source:* Robert W. Messler, Jr., *Principles of Welding,* Wiley-VCH, 1999, Figures 13.19 and 13.21, pp. 414 and 415; used with permission.)

Without going into detail here, as the purpose of this book is not to teach solidification but, rather, to teach problem-solving techniques, here using *cases,* Figures 15–2 through 15–4 depict the situation for each case (Ref. Messler).

In real welds, for example, there is tremendous turbulence (i.e., stirring) in the weld pool from a variety of forces. This would lead one to expect real welds to solidify according to Case 2. But fluid dynamics demands a stagnant boundary layer at the edge of the turbulent pool, where it abuts the unmelted base metal. Within this layer, Case 3 actually comes closer to representing reality.

In Summary

It is always important for an engineer to know the underlying givens, premises, or suppositions for a problem or an approach to solving a problem, as these define the *assumptions* that make the problem solution valid. Likewise, it is important for an engineer to recognize and operate within externally, internally, or self-imposed restrictions or limits, as *constraints,* as these fix the range of applicability of the solution to the problem. The prevailing circumstances from the standpoint of *initial conditions* and/or of *boundary conditions* are also important, with the latter being essential to finding solutions for problems involving differential equations. Two especially important and common conditions in engineering are (1) *equilibrium versus nonequilibrium* and (2) *steady state versus non–steady state versus transients.* Finally, it is sometimes useful for engineers to consider the set of circumstances or state and/or conditions that provide an example that can be precisely defined and/or calculated for use as a model or reference against which a problem for which a precise solution may be impractical or intractable can be compared. These are called *cases* and are analogous to those used by doctors in medicine, lawyers in law, and businesspersons in business.

Suggested Resource

Messler, Robert W., Jr., *Principles of Welding: Processes, Physics, Chemistry, and Metallurgy,* John Wiley & Sons, New York, 1999, pp. 402–422.

PART TWO

Physical/Mechanical Approaches to Problem-Solving

CHAPTER 16

Reverse Engineering

GALEN (AELIUS GALENUS OR CLAUDIUS GALENUS), the great Ancient Roman physician and surgeon (AD 129–c.200), did it to dramatically advance early medicine. Leonardo da Vinci (Leonardo di ser Pieor da Vinci, April 15, 1452–May 2, 1519), the genius mathematician, painter, sculptor, architect, musician, scientist, engineer, inventor, anatomist, geologist, cartographer, botanist, writer from Florence (later living and working in Rome, Bologna, Venice, and France), did it to learn from and mimic Nature. Paleontologists did it to learn what dinosaurs were and, later, deduce they were actually warm—not cold—blooded ancestors of birds. Anthropologists and archeologists did it—and do it—to learn about our past as a species. Unscrupulous "pirates" do it to illegally copy software. And, most important to the purposes of this book, engineers do it for a host of reasons, mostly perfectly legal and ethical. All employed (or employ) what is commonly known as reverse engineering. *Reverse engineering* is the process of or for discovering the technological principles underlying and enabling a human-made[1] or natural device, object, or system through systematic visual analysis of its structure, function, and/or operation.[2] In this sense, it is very much a problem-solving technique.

In most (but not all) instances, reverse engineering involves carefully taking a mechanical device, electronic component, biological entity, chemical or organic material or substance, or, more recently, software program apart and using deductive reasoning to analyze its workings in great detail. In this sense, it is the opposite of more usual "forward engineering," which involves synthesis versus dissection, and refinement and execution of a design concept versus abstract deduction of an original concept from the end product (Figure 16–1). For biological objects or systems, the process may involve dissection. In engineering, the process, by analogy, can appropriately be referred to as *mechanical dissection.*

Regardless of what it might be called or the physical object, device, material, substance, or system for which it may be employed, the reason or reasons for its use are many and varied, legal and illegal, ethical and unethical, scrupulous and unscrupulous, appropriate and inappropriate. Without too much comment on legality, ethicalness, scrupulousness, or appropriateness, major reasons for employing reverse engineering include:

1. *Design/product/production analysis:* To examine how a design, product, or production method works; what it consists of in terms of components, parts, or steps; what materials and/or processing methods seem to have been used; and how much it is likely to cost. This may be done as a normal part of seeking to continuously improve quality or performance, to advance a product or process, to improve efficiency and/or reduce cost, or to see if someone else's product or process infringes upon your company's patent rights.

---

[1]There are some who would include nonhuman—alien—made objects, such as UFOs, as well. Whether this has been done or not is subject to debate. But if any intelligent alien species ever visits or sends anything to Earth, we humans should—and almost certainly will—attempt to reverse engineer what they come in or send.

[2]In the case of case of paleontologists, anthropologists, and archeologists, they look at pieces of things to attempt to understand that thing when reconstructed.

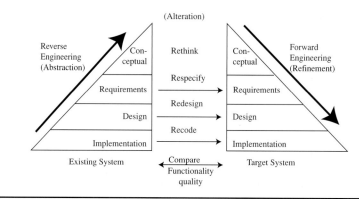

Figure 16-1 A schematic summarizing the comparison between the physical/mechanical problem-solving technique of reverse engineering versus more usual forward engineering.

2. *Competitive technical intelligence gathering:* To understand what, how, and why a competitor is doing compared to what, how, and why you are doing what you're doing or what, how, or why they say they are doing it. This is especially important in business.

3. *Learning:* To learn from others' or your own mistakes so as not to repeat those same mistakes. This is especially important in design.

4. *Troubleshooting inoperability:* To determine why something is not working as it should or once did. This can be important in design or manufacturing.

5. *Overcoming lost methods or documentation:* Sometimes knowledge of how a particular device or system or process, inanimate or animate, worked was lost or was never really known. Other times, the documentation for how (and, perhaps, why) a particular device or system or process was produced and operated has been lost over time or was never written down, and, to make matters worse, those who built it or operated it are no longer available (i.e., are dead or out of business).[3]

6. *Acquiring sensitive data:* To acquire data that were (or are) militarily confidential or industrially proprietary, hardware or software systems can be disassembled and the components analyzed to assess the design and its intent.

7. *Military or commercial espionage:* To learn about an enemy's or competitor's technological status by acquiring (by stealing, capturing, or fortuitous circumstances, such as a crash) a prototype and dismantling it. This is done all the time by sovereign nations and by smart companies as part of always knowing (and never under-estimating) your competition.

8. *Digital updating:* To create a digital version (e.g., a 3D model) of an object manufactured prior to computer-aided design (CAD) and manufacturing (CAM) or to update (and, possibly, correct) an earlier digital version.

9. *Recycling:* To sort and/or harvest materials from scrapped hardware for reuse in some form.

10. *Removal of copy protection:* To attempt to circumvent access restrictions.

11. *Creation of unlicensed/unapproved duplicates:* It should go without saying, this is generally not right, but is, without question, done. "Pirated" software and "knockoff" products are a serious breach of ethical business practice, not to mention the illegality. An exception can be "cloning," as occurred when many new companies entered the personal computer (PC) business by copying IBM's PC.

---

[3]The U.S. Navy implemented its Rapid Acquisition of Manufactured Parts (RAMP) program in the late 1980s. As part of this program, it needed a way to reverse engineer parts of ships made by companies no longer in business, often requiring the reconstruction of data for a new part from a severely worn or broken part. The analogy is a program opticians have for reproducing broken eyeglass lenses from only the broken pieces or some portion of those pieces.

12. *Curiosity:* To take something apart because it can be taken apart. This is how many, if not most, engineers-to-be got turned on to engineering: They took apart their toys or their mom's vacuum cleaner.

When it comes to machines or other products, for example, reverse engineering is commonly done as part of what is known as *value engineering.* In value engineering, products are routinely deconstructed (i.e., mechanically dissected) and analyzed with the purpose of finding opportunities for and avenues to cutting costs. The other common purpose of reverse engineering as businesses transition from old-fashioned blueprints to CAD drawings or CAD models (and data files for use in CAM) is to make 3D digital records of these products and/or to assess competitors' products.

As far as legality goes, in the United States even if an artifact (i.e., a physical object) or process is protected by trade secrets, reverse engineering the artifact or process is usually lawful as long as the artifact or process is obtained legitimately. For example, there is nothing illegal about purchasing a competitor's latest product on the open market and dissecting it. Likewise, there is nothing illegal in trying to duplicate Coca-Cola, which has been attempted over and over.

As for the question of reverse engineering being ethical, that comes down to intent. If, for example, the intent is to learn what can be done to improve one's own product, that is generally ethical. If the intent is to shortcut the effort to create one's own product or, worse, to deceive customers into thinking they have an original (e.g., Rolex watch) and not a knockoff, that is unethical!

Reverse engineering, as a formal process, actually had its roots as a tool in military applications. While surely used for centuries earlier, modern reverse engineering was often used in the Second World War (WWII) and in the Cold War between the United States and the Soviet Union (U.S.S.R.). Militaries of one nation routinely assessed, and often copied, other nations' technologies, devices, or information as these were obtained by regular troops in the field or by intelligence-gathering operations, or by one nation's good luck and another's bad luck with a crash or a defection. Some well-known examples are:

- In World War II, British and American forces noticed that the Germans had gasoline cans with an excellent design. They reverse engineered those cans to create what were popularly called "jerry cans" (Figure 16–2).
- The Soviets developed their Tupelov Tu-4 strategic bomber, not having had one before, by reverse engineering three American B-29 Superfortress bombers that had been forced to land in Siberia while on a mission over Japan during WWII. In just a couple of years, they developed a near-perfect copy (Figure 16–3).
- Both the Soviet Union and the United States got a jump-start on their respective rocket programs by reverse engineering the German V2 rocket that reeked havoc on London during World War II. Both the Soviets and the Americans captured technical documents, as well as V2 parts but, more important, German engineers, to create their R-1 and Redstone rockets, respectively. These turned out to be world-changing, as they paved the way for both intercontinental ballistic missiles (ICBMs) and the space race, the latter of which was won by the United States' landing on the moon on Monday, July 20, 1969. Photographs of a captured German V2 and U.S. Redstone are shown in Figure 16–4, while video of the Soviets using captured German V2s and scientists to create their RE-1 can be found on YouTube under "Soviet R-1 Missile."

Far less well known was the capture of a "superweapon" before it could be used by Japan against the United States in World War II. Rather than pursue an atomic bomb like the United States and Germany were pursuing, Japan decided to build a super-submarine capable of carrying aircraft that could—and were to—attack Los Angeles, New York, and Washington, D.C. The Japanese design overcame the problem of stability against rolling over due to being top-heavy by creating a side-by-side double-cylinder hull. After capturing the Japanese prototype, the United States immediately reverse engineered the design for what turned out to become the Trident missile-launching nuclear submarine. To keep the

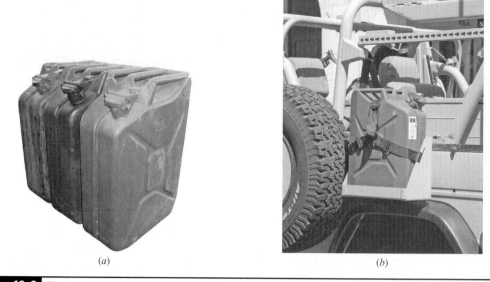

(a)   (b)

Figure 16-2 The jerry can, which was reverse engineered from examples seen and obtained from the Germans by British and U.S. forces during World War II. The original design (*a*) was copied for its simplicity and utility, and is still used by most military forces (*b*). (*Source:* Wikimedia Commons; [*a*] originally contributed by JohnM on 13 May 2006 and released to the public domain; [*b*] originally by DocteurCosmos on 8 May 2008 and used under GNU Free Documentation 1.2 or higher free license.)

secret from the Soviets, who knew of the Americans' good luck, the U.S. Navy blew up and sank the Japanese prototype in deep water off the California coast.

Whatever the purpose, reverse engineering is an important and powerful physical/mechanical approach for solving engineering problems. Like other physical or mechanical approaches, it also tends to be a favorite of many engineers, who seem to have a natural affinity for taking things apart and putting them back together.

(a)   (b)

Figure 16-3 Photographs of the American B-29 Superfortress strategic bomber (*a*) reverse engineered by the Soviets from three U.S. aircraft forced to land during a mission over Siberia to Japan during World War II. The result was the Tupelov Tu-4, a near-perfect copy (*b*). (*Source:* Wikimedia Commons; originally contributed by Rottweiler on 23 December 2006 and Ntmo on 11 May 2007, both used under GNU Free Documentation License 1.2 or higher. The photo of the B29 was taken by a U.S. airman in the line of duty, so is also part of the public domain as property of the U.S federal government.)

(a)   (b)

Figure 16-4 Photographs of (*a*) an American captured German V2 rocket fired from White Sands, New Mexico, in the early 1950s; (*b*) U.S. Army Redstone missile launched from Cape Canaveral, Florida, in 1958. Go to YouTube, under a search for "Soviet R-1 Missile," to see video of captured German V2s and scientists working on the Soviet's own R-1 missile. The latter two were reverse engineered from the original German rocket. (*Sources:* [*a*] From NASA files, for which no permission is required for use; [*b*] from official U.S. Army files, for which no permission is required for use; and [*c*] from www.daviddarling.info/images/R-1.jpg, with permission to use.)

To offer illustrative examples of reverse engineering is both daunting—for the detail required to execute the technique properly—and far less effective than the real thing, that is, with hands and eyes on. To "dissect" an automatic electric coffeemaker into a dozen parts and analyze each part's shape, role, connection to other parts, material of construction, and method of manufacture would take time—and lots of pages. To do this for a Jet Ski would take a long chapter or a small book. To do it for a captured V2 would take volumes. This notwithstanding, a couple of simple examples are provided to give the reader the essence of the technique.

Illustrative Example 16-1: A Circular Saw Blade

Your company decides it wants to add circular saw blades for electric-powered circular saws to its line of replacement disposable parts for tools and lawn and gardening

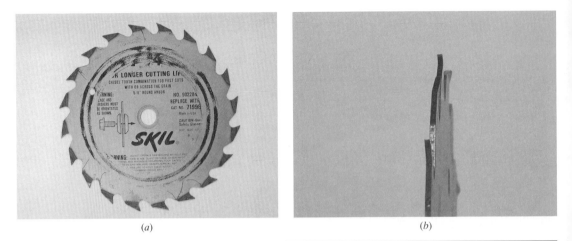

(a)   (b)

Figure 16-5   Photographs showing the blade from a Skil electric-powered circular saw. Note the discoloration of teeth from oxidation during induction heating of the teeth to allow hardening by water quenching (b). Also note how alternating teeth are slightly bent in and out of the plane of the blade body to produce a cut that is wider than the blade to prevent seizing. (*Source:* Photograph by Don Van Steele for the author. Owned by the author.)

equipment. Having never designed or produced saw blades before, your purchasing agent secures some electric-powered circular saws made by several manufacturers, one being Skil, a well-known and highly regarded manufacturer of power tools. You are assigned the job of reverse engineering the blades to decide on the best design for your product line.

Figure 16-5 shows one of Skil's blades. One of the first things you check is the material used to make the blade. You find it is magnetic, indicating it is a steel of some type and has a Rockwell C hardness of 28–32 measured at several locations on the face from the center outward toward the teeth. The teeth are raked in the direction of cutting, as indicated by stenciled arrows on the face of the blade. You note that the teeth are sharpened on their leading edges and that alternating teeth are slightly bent in opposite directions from the plane of the blade. You deduce that this is to cause the cut (i.e., *kerf*) to be wider than the blade body to prevent seizing. You decide the general shape—a round flat disk with unsharpened teeth—could be blanked from sheet-gauge steel, and the center hole could be punched in the same operation. A second operation could bend the teeth slightly inward or outward from the plane of the blade body, alternately, but probably not until the teeth were sharpened while all in plane. Checking the hardness of the teeth, you find they measure Rockwell C 58–62. A blue-black discoloration on the teeth suggests they were heated (into the austenite region) and quenched (to produce hard martensite) after they were sharpened and bent. Not seeing any indication of welds at the base of the teeth, you deduce that heating was done using a localized heat source, perhaps using rapid high-frequency induction, followed by water quenching. To reduce brittleness, a second light heating operation may have followed to temper the as-quenched martensite slightly. This approach created hard, wear-resistant teeth with a softer, tougher (i.e., more impact resistant) blade body.[4]

---

[4]Information on the steel used for the blade body and the steel used for the teeth could be obtained by reverse engineering either by conducting chemical analysis or, more simply (but approximately, but probably sufficiently), by polishing and 2% nital reagent etching samples of material from each area and looking under an optical microscope at 100X. The blade will likely consist of soft-ductile white-etching ferrite containing less than 022 wt.% C and dark etching hard pearlite containing 0.77 wt.% C. By using the "lever rule" (which can be found on the Internet or in a textbook on basic materials), the approximate carbon content can be determined from estimates of the proportions of each microstructural constituent present. The carbon content of the teeth, which

Looking at blades by other manufacturers, you find that some seem to have welded teeth onto the circular disk. In this approach, a prehardened narrow ring of steel would likely have been welded to the softer blank and then the teeth produced by grinding.

Illustrative Example 16–2: A Toddler's Toy

Pool toys for toddlers are made entirely of plastic and are totally free of metal parts and fasteners. This makes sense, as it precludes sharp parts and choke hazards that could come loose. Many such toys are inflatable, and some are made from Styrofoam, such as "noodles." The interested reader is encouraged to search online, finding images of several toys that are of different designs, in order to reverse engineer each mentally, as opposed to via mechanical dissection. Try it. It's fun. The more carefully you look, the more you see.

In Summary

Reverse engineering involves the mechanical dissection of physical artifacts for the purpose of learning how they work. The specific motivation can vary, but the approach is always the same—meticulous disassembly and vigilant observation—to see all there is to see.

Suggested Resources

Book
Eilam, Eldad, *Reversing: Secrets of Reverse Engineering,* John Wiley & Sons, Hoboken, NJ, 2005.

Website
"What Is Reverse Engineering?" on http://www.upd.solutions.com/reoverview.html.

---

will almost certainly be quenched martensite, can be approximated from the hardness alone (using plots of hardness versus C content in martensite available on the Internet).

CHAPTER 17

Material Property Correlations

THERE ARE TIMES when engineers seek to solve problems on the way to a design, for example, that involve finding a particular property of a material used in some part of the design. The property[1] may be a mechanical property (e.g., strength or stiffness), an electrical property (e.g., resistivity or conductivity), a thermal property (e.g., melting point or thermal conductivity), an optical property (e.g., transparency or opacity), a magnetic property (e.g., coercive strength), a chemical property (e.g., solubility or resistance to corrosion), or even a physical property (e.g., density). A good engineer would routinely seek the needed property from some reference source (e.g., a handbook or the Internet). But sometimes the needed property is not available. Then what? The answer resides in the technique of using *material property correlations.*

The most basic properties of materials (melting point, density, fracture strength, stiffness, electrical conductivity, thermal conductivity, specific heat, and others) have their origin in the atomic structure (or, for polymers, the molecular structure) of the material.[2] As a consequence, there are correlations between certain of these properties, often, but not always, taken two at a time. Whenever such correlations exist, they create the basis for data checking and estimating (see Chapter 6), and they often employ ratios (see Chapter 8). Hence, the value to problem-solving is obvious.

What is not so obvious, perhaps, is why *material property correlations,* whatever they may be, should be grouped as a problem-solving technique under physical/mechanical approaches, as opposed to as a technique grouped under mathematical approaches. After all, correlations are made through ratios. In the author's opinion, there are two good reasons: First, the use of ratios is incidental to the correlation between the properties themselves, as the latter has its basis in the physics of the system (i.e., the material). Second, materials are physical—not abstract—entities. Mathematics is abstract, not physical. Regardless of where material property correlations are listed, the technique can be very valuable in problem-solving.

To understand the basis of *material property correlations,* one only needs to look at the binding energy curves for solids for several good examples. Figure 17–1 gives schematic plots of the attractive, repulsive, and net force curves (*a*) and corresponding attractive, repulsive, and net potential energy curves (*b*) associated with atomic-level bonding. As two atoms (of the same or different elements) approach one another from infinite separation, at which distance of separation there is no force acting between them, they induce dipoles in one another by causing one another's negative charge center for orbiting electrons to shift slightly from one another's positive charge center for nuclei.[3] The force of attraction increases as the distance of separation decreases, as shown by the dashed curve for $F_A(x)$

---

[1] A *property* is a material's response to an external stimulus. Mechanical properties are the response to applied forces or loads. Electrical properties are the response to applied electromagnetic fields. Thermal properties are the response to heat. Optical, magnetic, and chemical properties are the response to light, an applied magnetic field, or a chemical agent, respectively.

[2] The fundamental principle underlying material science is that structure determines properties.

[3] In fact, there is also some force, even for atoms at infinite separation, albeit infinitesimal. Also, positive and negative ions (i.e., cations and anions) have ever-present and opposite electrical charges and, so, attract one another in ionic bonding. Likewise, polar molecules have permanent charges, so opposite-charged ends of approaching polar molecules attract to form one type of van der Waals secondary bonds.

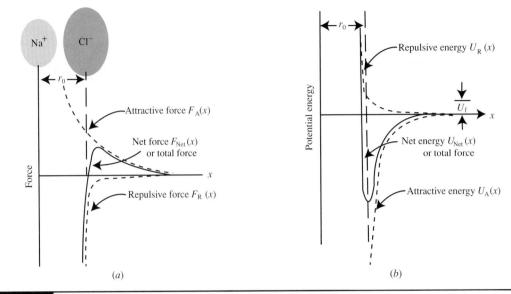

Figure 17-1   Schematic plots of the attractive, repulsive, and net force (*a*) and corresponding repulsive, attractive, and net potential energy curves for two approaching atoms (*b*).

in Figure 17–1*a*. As the atoms get very close together, their outermost (valence) electron shells begin to sense one another and a repulsive force develops that increases even more rapidly with decreasing separation distance, as shown by the dashed curve for $F_R(x)$ in this same figure. At some separation distance, the attractive and repulsive forces are equal, so they completely offset one another to result in a zero force in the net force, as shown by the solid $F_{NET}(x)$ curve in Figure 17–1*a*. Where the net force is zero, the two atoms are in equilibrium and form a bond.

Corresponding to force curves are curves for potential energy, as shown in Figure 17–1*b*. In the case of potential energy, the sign (i.e., direction) of the energy is opposite that of the force, but, nonetheless, attractive and repulsive potential energy components (associated with repulsive and attractive forces, respectively) sum to produce a net potential energy curve.[4] These are shown by $U_A(x)$ and $U_R(x)$ dashed curves and a U_{NET} solid curve in Figure 17–1*b*. Since Nature—in essence, through thermodynamics—seeks to minimize the energy of a system, the pair of atoms reaches a stable equilibrium separation at the point where the net potential energy is a minimum and the corresponding net force is zero.[5] At this *equilibrium separation distance,* the atoms are bonded.

Several features of the net potential energy curve for bonded atoms are the origin of certain properties of a solid aggregate of these atoms—that is, a solid material. These features and associated properties are:

1. The depth of the "well" in the net potential energy curve represents the energy it would take to separate the bonded atoms (i.e., the binding energy or bond energy). Thus, it gives rise to:

    a. The melting temperature or melting point (MP) for the solid material, as it would take this amount of thermal energy to break the bond and cause the solid to transform to a liquid. A deeper well indicates a higher MP material.

    b. The cohesive strength (or fracture strength) of the solid, as it would take this amount of mechanical energy to break the bond and fracture the solid. A deeper well indicates a stronger material.

[4]The relationship between F and U is: $F = -dU/dx$.
[5]The actual thermodynamic driver for bonding is the attainment of minimum potential energy.

2. The sharpness of the well in the net potential energy curve represents the stiffness of the bond (imagine it to be a spring) and, thus, the stiffness (as the modulus of elasticity) of the material.[6] A sharper, narrower well gives rise to a greater stiffness and higher modulus.

3. Finally, the symmetry (or, more correctly, the asymmetry) of the well in the net potential energy curve gives rise to the coefficient of thermal expansion of the solid material, as greater asymmetry causes the midpoint for an atom oscillating about a lattice point to shift outward as temperature rises from absolute zero, which the solid curve represents. Greater asymmetry gives rise to a larger value for the coefficient of thermal expansion (CTE).

Based on the net potential energy curve for bonding, and other atomic-level effects and interactions (e.g., electrons free to contribute to conduction), there are strong (but not perfect) correlations between the following material properties:

- Tensile strength (most correctly, fracture strength in tension) and melting point.
- Modulus of elasticity and melting point.
- Modulus of elasticity and tensile strength.
- Hardness (as a complex combination of strength, modulus, and ductility and/or toughness) and tensile strength.
- Density and specific heat (which arises from the ability to store vibrational, i.e., phonon, energy, which becomes easier as atoms are separated by greater equilibrium interatomic distances, as bond strength tends to weaken). Lower density favors higher specific heat.
- Thermal conductivity and electrical conductivity (as, in metals, both have their origin in the availability of free conduction electrons).
- Thermal expansion coefficient and melting point (as both relate to the sharpness and asymmetry of the well in the net potential energy curve).

Mathematically, the correlation between two material properties, P_1 and P_2, can be expressed as $C_L \leq P_1^m P_2^n \leq C_H$, where m and n are powers (usually, $-1/2$, -1, $+1/2$, or $+1$) and C_L and C_H are dimensionless lower and upper limits of the constant correlation factor C between which the value of the property group lies.[7]

Graphically (see Part 3), material property correlations can be seen in the distinct pattern or relationship exhibited by data points representing various materials in a plot (see Chapter 26, Graphing). Figures 17–2 and 17–3 give two examples for the material property data contained in Tables 17–1 and 17–2, respectively. Both plots show distinct patterns in the data.

The correlation between certain material properties is also the basis for material selection charts, as popularized by Michael F. Ashby (Ref. Ashby; see Figure 9.2 in this book, reprinted with permission from the original in Ashby), in which higher modulus E tends to occur in materials with higher density ρ).

Illustrative Example 17–1: Using Material Property Correlations

As an engineer, you are faced with a problem for which you need a reasonable value for the specific heat for zinc alloys. Unfortunately, you have not been able to find any values in available reference sources. Recalling there is a correlation between the specific heat of a solid and its density, you decide to employ this material property correlation to estimate a value of specific heat for zinc alloys.

---

[6] $F = -dU_{NET}(x)/dx$. Hence, $F = 0$ at the minimum of $U_{NET}(x)$. $dF_{NET}(x)/dx$ at the point x where $F = 0$ is proportional to the stiffness; i.e., $E \propto dF/dx$ or d^2U/dx^2.

[7] Values of C_L and C_H are not actually dimensionless (see Chapter 14) but, rather, have dimensions that are meaningless.

TABLE 17-1 Tensile Strength and Melting Point for Some Pure Metals

Metal	Tensile Strength* (MPa)	Melting Point (K)
Indium (In)	388	430
Tin (Sn)	11	505
Lead (Pb)	12	600
Zinc (Zn)	150	692
Magnesium (Mg)	275	923
Aluminum (Al)	90	933
Silver (Ag)	170	1234
Gold (Au)	100	1337
Copper (Cu)	220	1358
Nickel (Ni)	195	1726
Iron (Fe)	350	1808
Titanium (Ti)	240	1930
Platinum (Pt)	145	2045
Zirconium (Zr)	207	2125
Iridium (Ir)	585	2727
Molybdenum (Mo)	500	2890
Tantalum (Ta)	165	3269
Tungsten (W)	760	3683

*"Tensile strength" refers to the *ultimate tensile strength* for metals.

Tensile Strength versus Melting Point

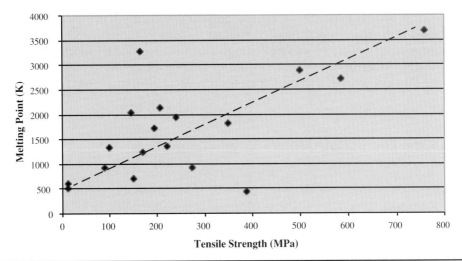

Figure 17-2 A plot of (ultimate) tensile strength versus melting point for pure metals to show the strong correlation between these material properties, which has its origin in the bonding energy curve.

TABLE 17-2 Thermal Conductivity and Electrical Conductivity for Some Metals and Alloys

Metal or Alloy	Thermal Conductivity (W/m-s)	Electrical Conductivity $(\Omega\text{-m})^{-1}$
Carbon steel	51.9	6.25×10^6
Austenitic stainless steel	16.2	1.39×10^6
Gray cast iron	46.0	0.67×10^6
Pure aluminum	222	34.5×10^6
7075-T6 Al alloy	130	19.2×10^6
Pure copper	388	58.1×10^6
70Cu-30Zn brass	120	16.1×10^6
Pure titanium	16	2.38×10^6
Ti-6Al-4V alloy	67	0.58×10^6
Pure gold	315	42.6×10^6
Pure silver	428	68×10^6
Pure molybdenum	142	19×10^6
Pure tungsten	155	18.9×10^6
Pure nickel	70	10.5×10^6
Inconel 625 Ni alloy	6.8	0.78×10^6
Pure lead	35	4.4×10^6
Pure tin	60.7	9.0×10^6
60Sn-40Pb solder	50	6.67×10^6

You find the density ρ of zinc alloys to be 7 to 8 g/cm³. You find the density of aluminum alloys to be lower, at 2.5 to 2.9 g/cm³, and for copper alloys to be higher, at 8.93 to 8.94 g/cm³, bounding the density of zinc alloys. The corresponding values of specific heat C_p for aluminum alloys and copper alloys are 875 to 963 J/kg-K and 375 to 420 J/kg-K, respectively.

Using the relationship that $C_L \leq \rho C_p \leq C_H$, and the knowledge that lower density favors higher specific heat, you estimate values of C_p for zinc alloys thus:

For Al alloys: (2.5)(963) ~ 2410 and (2.9)(875) ~ 2540
For Cu alloys: (8.93)(420) ~ 3750 and (8.94)(375) ~ 3350

The narrowest range is thus 2540 to 3350, yielding approximate values of the specific heat C_p for zinc alloys (with densities of 7 to 8 g/cm³) of 363 to 475. This range corresponds quite well to the actual value of C_p for pure zinc of 395 W/kg-K. Moreover, the most popular zinc-based alloys have small amounts (3 to 6 wt.%) of aluminum, which lowers the density somewhat, to slightly increase the value of C_p somewhat above 395. Close enough when one adheres to the author's premise that one should never measure a manure pile with a micrometer.

In Summary

The fact that certain basic properties of materials are related through atomic-level effects and/or interactions gives rise to correlations between some properties. Such *material property correlations*

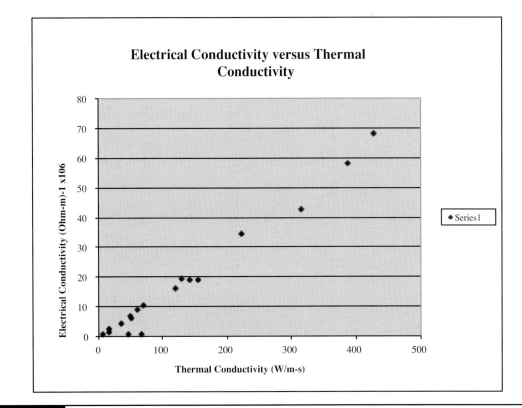

Figure 17-3   A plot of thermal conductivity versus electrical conductivity for some metals and alloys to show the strong correlation between these two properties, which has its origin in both arising, in metals, from the availability of free conduction electrons.

can be used in problem-solving to estimate a property for which data cannot be found from data for a correlated property for which data can be found.

Suggested Resources

Ashby, Michael F., *Material Selection in Mechanical Design*, 3rd edition, Butterworth-Heinemann/ Elsevier, New York, 2005.

Messler, Robert W., Jr., *The Essence of Materials for Engineers*, Jones & Bartlett Learning, Sudbury, MA, 2010.

CHAPTER 18

Proof-of-Concept Models

MODELS ARE PHYSICAL embodiments of ideas and/or designs. They are invaluable to engineers as they engage in problem-solving at all stages, from concept generation to first-off prototypes. The reason physical models are so important and useful is that, as physical beings living in a physical world, physical objects or embodiments are easier for us to understand, evaluate, and react to than abstract ideas or entities.

Corresponding to the various stages of idea development and problem-solving, the author holds that there are several types or categories of models by which evaluation can be done. These are described and discussed in this and the next four chapters, as follows, in order of increasing sophistication and from earliest to last stage of use:

- Proof-of-concept models (Chapter 18)
- Experimental models (Chapter 19)
- Test models (Chapter 20)
- Mock-ups and masters (Chapter 21)
- Prototypes (Chapter 22)

A *proof-of-concept* (POC)[1] or a *proof-of-principle model* is a physical embodiment (or *realization*) of an idea or a certain method intended to show that idea or method's feasibility. In some cases, it demonstrates some concept or principle to verify that the concept or principle has potential, if not particular, value for use alone or as part of a more complicated solution or design. In most cases, a proof-of-concept model is small and, whenever possible, simple, and it may or may not represent an overall concept, solution, or design (i.e., it may represent only a portion of the overall concept, solution, or design).

For the author, as both an engineer (who happens to teach) and as a teacher of design, a proof-of-concept model is most valuable in problem-solving to prove that a potential "technical showstopper" doesn't exist or, if it does, can be overcome on the way to the final problem solution or design. A "technical showstopper" is that aspect of a design concept or potential solution to a problem that would prevent the design from working or the solution from being viable. Recognizing potential showstoppers is *much* easier with experience, and potential showstoppers are difficult to describe in a book as forewarnings for young engineers. This said, an engineer (or engineering student) should mentally run through a design concept, a new process, or, in searching for the solution to a problem, the tentative solution, to see if there is something—anything—that would obviate the design, process, or solution. As but one example: In working on a spacecraft to carry human beings into space, the showstopper would—or should—be: Can they be brought back safely? Getting them into space is only part of the problem, and not the most important part, in the long run.

Proofs of concept are used in fields other than engineering, including:

- In filmmaking (e.g., to test the concept of a bluescreen before using it to make a movie)
- In security (e.g., to show whether a particular concept will provide protection and not be able to be compromised, without building an entire system)

[1]The term *proof of concept* seems to have first appeared around 1973, although this is uncertain. Some, like Bruce Carsten, take credit for it much later (i.e., 1984), saying he was the first to use it in his column "Carsten's Corner." Sort of like former vice president Al Gore saying he invented the Internet!

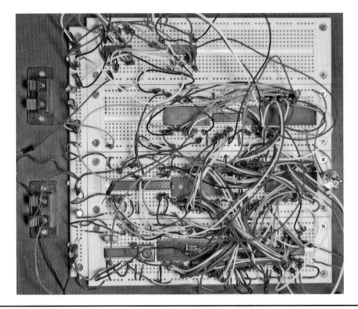

Figure 18-1   A photograph of a typical breadboard used by electrical engineers as a proof-of-concept model to prove a circuit works. (*Source:* http://www.todayandtomorrow.net/2006/04/24/the-breadboard-band; used with permission.)

- In software development (e.g., to try a new paradigm)
- In clinical drug development (e.g., to test effectiveness and safety)

However, the most frequent and diverse use is surely in engineering. Here, a rough model that embodies a key principle to be tested is usually used (e.g., to prove that a specific motion or mechanism actually works). In electronics, electrical engineers use what are commonly known as breadboards. A *breadboard* (or, alternatively, a *protoboard*) is a construction base for a one-of-a-kind electronic circuit, often to test an idea. For simplicity, solderless breadboards known as *plugboards* are used (Figure 18–1). Such breadboards can be used for small analog or digital devices to complete central processing units (CPUs).

For mechanical systems, simple jury-rigged or kludged models are often used for proof-of-concept testing.[2] To keep costs down and to expedite testing, designs must be kept simple and readily available hardware should be used (i.e., adopted) wherever possible. In the example of the proof-of-concept model for a device of unknown purpose—and of unknown origin—in Figure 18–2, notice should be taken of the following ingenious facets of what appears to be a jury-rigged apparatus:

- The brake system from a bicycle (near the person's hand), probably to actuate a cable running through the tube
- A thick rubber band (bottom, left-center), probably being used for either attaching something or as a tensioning device or a spring device
- Wheels, for making the model portable.
- Use of wood (in the base), easy-to-machine Al alloy (lower left), standard Al-alloy extrusions (in some vertical members), and common clamps (top, left-center)

It was said that the design for folding wings to allow Navy aircraft to be stowed more efficiently on and belowdecks on World War II aircraft carriers was shown by an engineer at the Grumman Aircraft

---

[2]*Jury-rigged* is an adjective that means "done or made using whatever is available." *Kludged* means "a clumsy or inelegant—[crude]—[temporary] solution to a problem."

Figure 18-2 A photograph of a proof-of-concept model for a mechanical apparatus of unknown (and, in the context of this book, unimportant) purpose. Note the ingenious use of readily available bicycle hand brake, rubber bands, and low-cost materials. (*Source:* Unknown.)

Company (where the idea originated) to the chief engineer using carefully bent steel paper clips (to represent wings) stuck into a pink draftsman's eraser (to represent the fuselage). A simple rotation of the paper clips about the pivot point where they attached to the eraser moved the "wings" from "flying" position to "folded" position, like the wings of birds or beetles.

While not a requirement to obtain a U.S. patent, inventors often build proof-of-concept models (Figure 18–3).

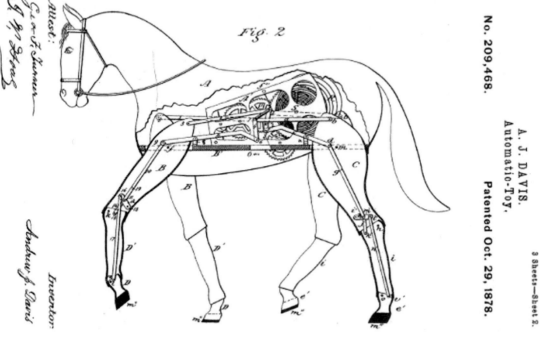

Figure 18-3 The drawing for a toy mechanical horse filed for patent by Andrew J. Davis on July 26, 1878, and issued a patent (U.S. Patent 209468) on October 29, 1878. Whether an actual proof-of-concept model was ever built by the inventor is unknown, as whether an actual toy was ever produced is unknown. (*Source:* U.S. Patent 209468, in the public domain.)

The term *brassboard* is occasionally used to refer to a proof-of-concept model for use in a mechanical device or system.

Illustrative Example 18–1: Need for a Proof-of-Concept Model

In a team-based senior capstone design project, a group of students was charged with finding various ways to change the shape of a wing for an airplane as part of an initiative to study morphing wing aircraft. The desire was to change as many geometric aspects of the wing as possible, including span (i.e., root-to-tip length), sweep angle (measured from the fore-to-aft centerline of the fuselage), wing width (i.e., leading edge to trailing edge), and chord (i.e., airfoil) shape. One team of mechanical engineers (MEs) mentally locked in on using a four-bar linkage mechanism (like what is found in expandable elevator gates), never seriously considering any other approach. The linkage would allow shear of the wing to cause sweep and/or could lengthen or shorten span (with an accompanying fixed amount of width increase or decrease, respectively). The plan was to use it as the substructure underlying a flexible thin elastomeric latex skin to provide streamlining.

The design became more and more elaborate (and potentially less reliable!) as the students tried to actuate the linkage using servomotors at each node or pivot point.[3] Worse yet, the team proceeded to build a *prototype* (see Chapter 22) without ever testing their concept to see if, as the author suggested as one of several faculty advisors, it might not be able to tolerate the stresses associated with bidirectional stretching of the overlaying skin so that it would always be taut, regardless of the wing shape. The expressed concern was the complication of straining under biaxial stress.

Sadly, during a final presentation of their prototype, the four-bar linkage buckled and failed, as the stress imposed on the linkage by the latex was too much!

All of this embarrassment could—and should—have been avoided by a proof-of-concept model to check out what was posed as a potential technical showstopper.

In Summary

Early in the process of solving a problem or creating a design, an engineer frequently needs to demonstrate—first to himself or herself, then to others—that some novel and yet-unproven concept or principle works. This is done using *proof-of-concept models* or *proof-of-principle models*. Such models should be as simple as possible, focus on the essential concept or principle, and not seek to be detailed, perfect, elegant, or aesthetically pleasing. Low cost and expediency on the way to proving a key point are the goals.

---

[3] No thought was given to what would happen if any servomotor failed. Even with some discussion by the students of redundancy—so one motor could pick up for another—no thought was given to what would happen if a motor seized, locking the linkage at that pivot point.

CHAPTER 19

Experimental Models and Model Experiments

BEFORE IDEAS BECOME reality in engineering, it is not uncommon for experimentation to be done. In fact, with modern use of solid modeling and computer-aided design, validation of a design with a physical model is more than a good idea, it is critical! After all, computer-based models are only as good as the programmer *and* programming, and they are virtual, not real. This is obviously even more true for revolutionary ideas or designs than for evolutionary ones.[1]

Experimentation in engineering, at least in the context of this book, takes two generic forms: (1) a structured set of carefully contrived and/or controlled conditions (as inputs) to observe effects (as outcomes), that is, *model experiments,* and (2) as physical embodiments or models to be proven as feasible, valid, and/or practical, that is, *experimental models.*

In the context of physical models being discussed (in Chapters 18 through 22), experimental models are most interesting. However, brief mention of and description of model experiments will also be made if for no other reason than to avoid confusion between the two similar terms in the chapter title.

Experimental models are most commonly found in engineering related to vehicles, being complex electromechanical systems, including experimental aircraft, experimental automobiles, experimental boats or ships, and experimental trains. All experimental models have in common that they are intended to prove some unproven idea. For experimental aircraft, the idea might be some still-unproven means of propulsion, specialized aerodynamics, enhanced maneuverability, novel control, or the like.

Here are two good examples: First is the concept for vertical/short takeoff or landing (V/STOL) aircraft, such as the LTV XC-142 (Figure 19–1).[2] The experimental model was intended to prove lift could be achieved and stable transition from vertical to horizontal flight could occur. Second, the concept for forward-swept wings, for the potential that such designs offered for exceptional maneuverability, was demonstrated by Grumman's X-29 for DARPA/NASA (Figure 19–2). The experimental model was intended to prove that stability could be achieved by computer control of flight surfaces, even though a forward-swept wing gives rise to inherent aerodynamic instability, and to prove that advanced composites could overcome problems associated with the severe stresses associated with divergence of forward-swept wings.[3]

Examples of experimental models of extremely fuel-efficient and "green" automobiles (winners of Shell Oil Company's Eco-Marathon)[4] and a biofuel powered "green" ship (e.g., Earthracer) are shown in Figures 19–3 and 19–4.

[1]*Revolutionary* ideas or designs tend to be significant, if not radical, departures from earlier ideas or designs, often involving new concepts or principles. Revolutionary ideas or designs often bear little resemblance to earlier ideas or designs. *Evolutionary* ideas or designs emerge as some form of improvement on an earlier idea or design, but the new idea or design is readily recognizable as an advance on the early idea or design.

[2]The "X" in the prefix for the designation of an aircraft denotes "experimental."

[3]To save money and expedite fabrication of two experimental aircraft, newly designed forward-swept wings made of advanced composite materials were attached to the fuselages of General Dynamics F-16s purchased as surplus. To do this, a new wing-carry-through structure had to also be added to the old aircraft.

[4]The Shell Oil Company sponsors an annual competition in various sectors of the world (the Americas, Europe, Asia) to allow inventors and inventive teams to demonstrate their experimental vehicles in various full categories (solar, hydrogen, etc.) and events. In 2007, a world-record mileage of 10,705 mpg (4551 km/L) was achieved with the French "Microjoule".

Figure 19-1 A photograph of the Ling-Tempco-Vought LTV XC-142A vertical/short takeoff and landing (V/STOL) aircraft during takeoff. The full-scale experimental model was intended to demonstrate stable lifting, hovering, and landing, as well as stable transition from vertical to horizontal flight and back again. (*Source:* Wikimedia Commons; NASA Photo ID EL-2001–00399 originally contributed by Kadellar on 30 December 2010, used for free under the public domain.)

Figure 19-2 A photograph of the Grumman Aerospace Corporation's X-29 forward-swept-wing fighter aircraft. The experimental model (of which there were two) was intended to demonstrate the ability to overcome inherent aerodynamic instability for such designs using computer control to make more than 50 corrections a second and to show that use of advanced composites in the wings provided sufficient stiffness to prevent twist from divergence of the wing tips. This test was at NASA's Ames-Dryden Flight Research Center, Edwards, CA. (*Source:* Wikipedia Commons; NASA Photo ID EC87–0182–14 originally contributed by Uwe W. on 20 December 2006, used for free under the public domain.)

Experimental models differ from *test models* (discussed in Chapter 20) in that the former are not close to production and, in fact, rarely make it into production. They are created to gain knowledge for a new design. Many times, experimental models are created for what will be a one-of-a-kind or small production volume. This is seldom, if ever, the case for test models. The latter are almost always early (if not first) production units for which full production is intended. They are sometimes pulled from production to certify performance.

Illustrative Example 19-1: An Experimental Model

In the late 1970s, the U.S. Department of Energy, shaken by the Arab oil embargo that drove prices high and created waiting lines at gas stations, aggressively pursued nuclear fusion as a safer alternative to fission as a long-term solution to the country's need for renewable energy. The Tokamak Fusion Test Reactor (TFTR) was an experimental tokamak-type device built and tested at Princeton's Plasma Physics Laboratory in Princeton, New

(a)   (b)

Figure 19-3 Photographs showing winners from two different classes of experimental automobiles in the 2011 Shell Eco-Marathon of Europe: (*a*) Nr 405 by a Team Politecnico di Milano, Italy, winner in the Urban Concept/Gasoline category; (*b*) Nr 502 by Team Lycee Louis Delage, France, winner in the Solar category. (*Source:* http://www.flickr.com/photos/shell _eco-marathon/collections/72157626374304149/; used with permission under license from the Shell International Ltd.)

Jersey (Figure 19-5). With a toroid plasma chamber with a major diameter (in plan view) of 4.2 to 6.2 m and a minor cross-sectional diameter of 0.8 to 1.92 m, a 6.0-Tesla magnetic field contained an 81-MW plasma to support fusion of deuterium and tritium.

The experimental model operated from 1982 to 1997 and, in the process, achieved the highest plasma temperature ever attained (at 510,000,000°C, with 110,000,000°C needed for fusion), the first "supershot" to produce many more thermal neutrons for producing

Figure 19-4 A photograph of a biofueled experimental model by Craig Loomis Design Group (CLD) of New Zealand. Intended to demonstrate the wave-cutting ability of the design, as well as the ability to use biofuel, "Earthracer" circumnavigated the world in 61 days. It is shown here as "Earthrace" in Hamburg, Germany. (*Source:* Wikimedia Commons; originally contributed by flickr on 28 November 2007, used under Creative Commons Attribution 2.0 Share Alike Generic free license.)

Figure 19-5 A photograph of the TFTR experimental model of a tokamak-type reactor for controlled nuclear fusion to produce electrical energy. (*Source:* Wikimedia Commons; originally contributed by Sheliak on 28 August 2009, used under Creative Commons Attribution 3.0 Unported free license.)

electricity, and a record 10.7 MW of fusion power (enough to supply 3000 homes with electricity) from a 50/50 mixture of deuterium and tritium in 1994.

The experiment also led to the design of subsequent experimental reactors of more advanced designs using, for example, sheared plasma.

The TFTR was an extraordinary experimental model in the pursuit of commercially viable consumer power from fusion reactions.[5]

While not a physical model, *model experiments* are important to some problem-solving, and, because of the possible confusion between the terms *model experiments* and *experimental models,* the technique will be very briefly described here.

Created by *design of experiment* (DOE) or *experimental design, model experiments are* information-gathering exercises intended to maximize the quantity and value of data from a minimum number of experiments. The technique involves the use of statistics in the design of the set or array of experiments to be conducted, in order to give the greatest statistical validity to the results obtained.

The interested reader is encouraged to consult any of several good references devoted to experimental design (see Suggested Resources).

A simple example of the use of model experiments is shown in Figure 23–2, under Illustrative Example 23–1, for using guided empiricism in a trial-and-error approach to problem-solving.

---

[5]The author, as a 30-year-old engineer and technical manager for advanced metallic structures, served as materials and processes coordinator for New Energy Ventures from 1977 to 1980, and, in that capacity, was liaison to Grumman engineers working at Princeton.

In Summary

Unproven ideas or concepts applied to create radical new designs must be proven to be technically feasible, practical, and operationally safe. This is often done using *experimental models.* While seemingly related (because of the titles), *model experiments,* which also are a powerful tool for problem-solving when experimentation is involved, are actually a statistically based method for designing the array of experiments that should be performed to maximize the utility and statistical validity of the data obtained from a minimum number of experiments.

Suggested Resources

Anthony, Jiju, *Design of Experiments for Engineers and Scientists,* Butterworth-Heinemann/Elsevier, New York, 2003.

Box, G.E.P., W. G. Hunter, and J. S. Hunter, *Statistics for Experimenters: An Introduction to Design, Data Analysis, and Model Building,* John Wiley & Sons, New York, 1978.

Montgomery, D. C., *Design and Analysis of Experiments,* 7th edition, John Wiley & Sons, Hoboken, NJ, 2009.

CHAPTER 20

Test Models and Model Testing

AT SOME POINT in the development of a new design for production, the actual or close-to finalized design must be tested to validate that the design works as planned and, sometimes, to assess and gather data on performance capability. For aircraft, for example, testing might be to assess takeoff, flight, and landing characteristics and performance. For high-performance automobiles, testing might be to assess acceleration, maneuverability, stability at speed, and braking. Whatever the myriad reasons, *test models* are invaluable as not only the final stage in completing a new design but, possibly, as the first stage for setting in motion the next-generation design. Test models differ from experimental models in that they are actually to be a production item, and their use is to check on, and perhaps certify, performance, not simply gather data on feasibility for eventual development and/or design.

As the chapter title suggests, models involved in testing for new planned products are of two types: (1) *test models* and (2) models for testing (i.e., for *model testing*). Both types differ from experimental models in that testing is to assess performance of designs involving known and/or proven concepts and technologies. These models are not intended to prove any unproven idea, approach, or the like.

Models for testing in *model testing* are almost always (but not always) subscale (versus full-sized). While model testing had unquestionably been done for centuries (if not millennia),[1] it came of age with efforts to create powered heavier-than-air flying machines. At the end of 1901, Wilbur and Orville Wright were frustrated by the flight tests of their gliders of 1900 and early 1901. Neither performed as well as predicted using available design and analysis methods. Lift, in particular, was one-third of what was predicted using available data for lift coefficients. To remedy this, the brothers built many small models of their wing designs (using materials available in their bicycle shop) and tested them in a wind tunnel of their own creation. They generated their own data on lift coefficients. Ever since then, scaled models of aircraft (or aircraft components, such as wings or fuselages), spacecraft, automobiles, and high-speed trains have been—and are—tested in wind tunnels in model testing (Figures 20–1 and 20–2).

The value of using small, subscale model testing in the process of problem-solving is obvious: lower cost for the model and for the testing facility.

There comes a time, however, when a scale model and model testing are not sufficient. Necessary, yes; sufficient, no. At some point before releasing a design for full production, full-scale test models are needed.

Test models are (or are intended to be) fully representative of the planned production item. In fact, some test organizations (e.g., Underwriters Laboratories, *Road and Track* magazine, or *Car and Driver* magazine) secure their "test models" from the marketplace. For planned production items requiring performance certification, however, while representative of the intended production item, the actual tested model is built to be tested and not to be sold. The Federal Aviation Administration, for example, uses test models specifically submitted by manufacturers for testing.

It is the rule, more than the exception, that aircraft manufacturers build three vehicles explicitly for testing: one for static testing, one for fatigue testing, and one for flight testing. The former is intended to check that all major structural members can carry the loads imposed by certain routine required flight maneuvers, as well as

[1] There is no doubt that Leonardo da Vinci tested some of his ideas as subscale models, although not all.

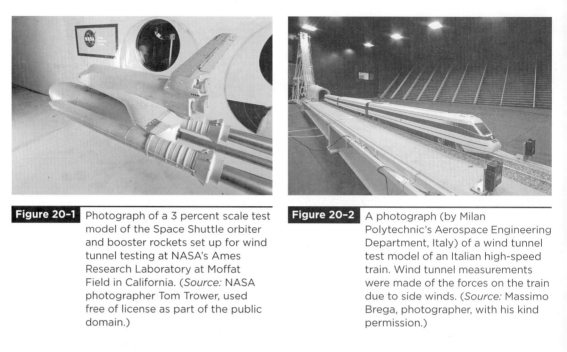

Figure 20-1 Photograph of a 3 percent scale test model of the Space Shuttle orbiter and booster rockets set up for wind tunnel testing at NASA's Ames Research Laboratory at Moffat Field in California. (*Source:* NASA photographer Tom Trower, used free of license as part of the public domain.)

Figure 20-2 A photograph (by Milan Polytechnic's Aerospace Engineering Department, Italy) of a wind tunnel test model of an Italian high-speed train. Wind tunnel measurements were made of the forces on the train due to side winds. (*Source:* Massimo Brega, photographer, with his kind permission.)

takeoffs and landings. For fighter aircraft, for example, static testing includes loads imposed by planned weapon stores, as well as tests to assess just how high static loads could go without loss of "safety of flight." The second model for fatigue testing is to assess needed and expected life, as affected by the innumerable cyclic loads incurred during flight, as well as takeoff and landing. These static and fatigue test aircraft are generally not flown but, rather, are subjected to loading in specially designed and computer-controlled fixtures. The third model is usually used for flight testing by a test pilot, ending with acceptance and certification of the aircraft by the military service or FAA in the United States.

Test pilots at aircraft companies and test drivers at automobile companies put a flying or driving test model through a predetermined set of sequential tests. Once these are completed, they often wring out the test model to see what it can do—and take. In civil aircraft flight-testing, tests are designed to verify that the aircraft meets or exceeds all applicable safety and performance requirements of the government certifying agency (the FAA in the United States and Canada). In military aircraft flight-testing, tests are designed to demonstrate that the aircraft can fulfill all mission requirements. In modern air-weapon acquisition, the air force and navy insist on "fly-before-buy" test model demonstrations and evaluations.

Besides testing performance against design specifications for users, test models are used to assess crashworthiness for operator and passenger survival. This is done for civil aircraft and for automobiles (Figure 20–3).

Illustrative Example 20-1: A Test Model

On July 20, 2011, the Space Shuttle *Atlantis* (ST-135) returned from a 12-day mission to resupply and make repairs and improvements to Orbiting International Space Station (Figure 20–4), bringing to an end the 30-year-long program that began in 1981. With no replacement agreed to by the U.S. Congress, the future of manned space exploration seemed uncertain.

The Space Launch System, or SLS, is a space shuttle–derived heavy launch vehicle designated by NASA to replace the retired Space Shuttle. The NASA Authorization Act of 2010 envisioned transformation of the Ares I and Ares V vehicle designs into a single

Figure 20-3 A photograph of a test of crashworthiness using a test model automobile; here, a 2010 Hyundai Tucson GLS being tested at the Insurance Institute for Highway Safety, Vehicle Research Center. (*Source:* Wikimedia Commons; originally contributed by Bpaulh on 7 August 2010 and used under Creative Commons Attribution 3.0 Unported, free license.)

Figure 20-4 A photograph of the Space Shuttle *Atlantis* docked at the Orbiting International Space Station during the last mission of the Space Shuttle on July 8–21, 2011. (*Source:* NASA; no permission for use is required, as it is in the public domain.)

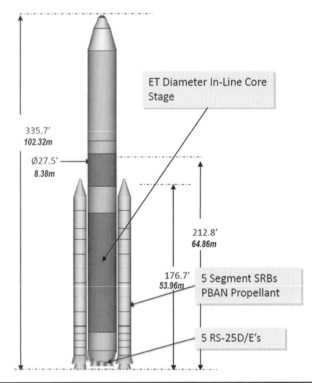

Figure 20-5 An illustration of NASA's SLS reference configuration. (*Source:* Wikimedia Commons and NASA, originally contributed by Haplochromis on 15 June 2011 and modified by Stanqo on 15 September 2011; used freely as part of the public domain.)

Figure 20-6 An artist's highly realistic, computer-rendered concept of the launch of an SLS. (*Source:* Wikimedia Commons and NASA, originally contributed by Uwe W on 15 September 2011; used freely as part of the public domain.)

Figure 20-7 Launch of the Ares 1-X test vehicle from Kennedy Space Center, Florida, on October 28, 2009. (*Source:* Wikimedia Commons and NASA, originally contributed by RTas67 on 28 October 2009; used freely as part of the public domain.)

launch vehicle usable for both crew and cargo. It is to be upgraded over time with more powerful versions. Its initial capability of core elements, without an upper stage, will be 70 metric tons (for the Block 0 configuration with three engines and a partially fueled core) and 100 metric tons (for the Block 1 configuration with four engines and a fully fueled core), both of which are suitable for low-Earth orbit (LEO) missions. Eventually, the Block 1 configuration will be upgraded to allow missions beyond low-Earth orbit. With the addition of an integrated Earth Departure Stage and a fifth SSME-derived core engine, the total lift capability is to be 130 metric tons.

Figure 20–5 is an illustration of NASA's SLS reference configuration, while Figure 20–6 is an artist's concept of the launch of and SLS. Figure 20–7 shows the Ares 1-X test vehicle during launch on October 28, 2009.

In Summary

Test models and *model testing* are essential tools (or techniques) for engineers to solve problems and, especially, to assess new designs. Test models are generally full-size units either planned as part of initial production or pulled from early production. Their intent is to certify that performance is what it is supposed to be. Model testing usually employs subscale models to assess specific performance characteristics to either validate design decisions or guide final design refinements.

CHAPTER 21

Mock-ups and Masters

TWO OTHER TYPES of models used in engineering to solve problems, almost always during design, are mockups and masters. A *mock-up* is a scale or full-size model of a design, device, assembly, structure, system, or, even, process intended for demonstration, design evaluation (including refinement and detailing), teaching, training, or promotion. They may be used to aid in either design or manufacturing. A mock-up becomes a *prototype* (see Chapter 22) if it provides all or a significant portion of the functionality of a system and, as such, enables testing of the design. Rarely do mock-ups provide any functionality. They are more about geometry. A *master* is usually a full-size model, but could be subscale for some purposes, used to assess a design for its appeal (i.e., aesthetics, including style, line, form, texture, etc.), and, often, as a pattern for extracting geometric and dimensional data to allow fabrication of the real object. As such, masters are more important to manufacturing than they are to design or testing.

Mock-ups are used virtually everywhere a new product is designed. Examples include:

- As an integral part of the acquisition process for the military and space agencies. The most common use is to test or assess human factors (i.e., ergometrics, such as pilot visibility, operator and passenger comfort, and accessibility). (See Figure 21–1.)
- In the aircraft industry to facilitate in the exact placement of wires, hydraulic lines, fuel lines, control linkages, heating and cooling ducting, and so on, which is still too difficult to do reliably using computer-based solid models (Figure 21–2). It may come as a surprise to young engineers-to-be, but physical mock-ups are widely still used in lieu of reliance on even the most sophisticated computer-based solid modeling (i.e., CAD). In civil aircraft, mock-ups are commonly used to allow assessment of alternative interior designs (Figure 21–3).
- In the automobile and automobile accessory industry as part of the product development process to assess ergometrics (human factors) and to test consumer reaction.
- For consumer products industries (electronics, appliances, furniture, etc.) to finalize dimensions, to assess human factors, and to gather overall reactions as part of market research.
- In architecture to assess visual appeal (aesthetics), client reaction, and fit with a site.

Mock-ups are also used in software engineering, principally to assess the human interface.

A common use of mock-ups is for teaching and/or training. For example, virtual reality alone can be used to train pilots in flight simulators, but for the pilots to get a physical feel for the aircraft or spacecraft itself, full-size mock-ups are commonly used.

Illustrative Example 21–1: Using Mockups

As design of the Lunar Excursion Module (LEM), intended to land a man on the Moon for the first time, progressed at Grumman Aerospace Corporation in the early 1960s, following President John F. Kennedy's challenge to Americans in a speech before a special joint

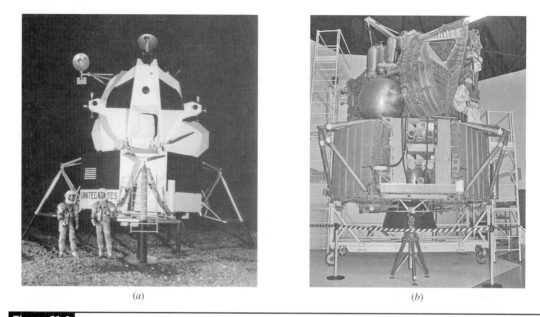

(a)   (b)

Figure 21-1 Photographs of two different wood mock-ups of the Lunar Excursion Module (LEM) built by Grumman Aerospace Corporation: (a) a photograph owned by the Cradle of Aviation Museum, Garden City, New York, of a display mock-up; (b) a photograph of LTA-1 used by Grumman to test different window configurations, flight-control actuators, and so on, with help from the astronauts. (*Source:* [a] Cradle of Aviation Museum, used with their permission; [b] photographed by Daniel L. Berek and used with his kind permission.)

Figure 21-2 Wood mock-up of the nose and cockpit of a Bombardier de Havilland Canada Dash 8 Q400 turboprop passenger airplane used for locating wiring. (*Source:* Photographed by Swire Chin and used with his kind permission.)

Figure 21-3 Photograph showing the interior layout for a Singapore Airline A380. (*Source:* http://www.aircarftinteriorsexpo.com, photograph taken by Aart van Bezooyen; used with his permission.)

session of Congress for America to safely land a man on the Moon and bring him back before the end of the decade, the time came for some details to be worked out. Three full-size plywood mock-ups of the LEM, designated LTA-1, -2, and -3, contained, among other design alternatives, three different window configurations. A group of 3 of the first 14 astronauts came to Bethpage, Long Island, New York, to visit Grumman and see the LEM for the first time. Their real purpose was to help design engineers decide, among other details, on a final design for the windows to afford maximum visibility.

While working at Grumman as a summer student each year through his undergraduate education at Rensselaer Polytechnic Institute (RPI), the author, following his sophomore year in 1963, waited anxiously with hundreds of other Grumman workers as astronauts Lt. John H. Glenn, Jr., Malcolm Scott Carpenter, and Lt. Col. James A. McDivitt climbed onto the realistic fully painted and outfitted mock-ups to sit in the pilots' seats and check out the windows. Young Lt. John Glenn, looking smart in his dress-blue Marine uniform, stopped and picked up the seat cushion from the mock-up and signed his name on the underside. When asked why he did that, he said, "I plan to fly to the Moon in a LEM one day soon. The only thing that will stop me is if I look under the seat cushion before we launch and I see my signature. In that case, I'll know it's still plywood."

Figure 21–4 shows the autographs of these American heroes obtained that day by an engineer-to-be.

Masters are used to assist in preparing what will be needed by manufacturing to allow production of hardware (e.g., vehicles). In the boat-building industry, "loftings" (see Chapter 25) are made, to serve as 2D templates to allow the construction or fabrication of a complex 3D shape, such as a hull. These 2D "slices" are taken from full- or, occasionally, subsize masters made from wood. In the automobile industry, subscale or full-size models made from sculpted clay or plaster or machined from Styrofoam are used by both the stylist (concerned more with aesthetics) and the designers (concerned more with functionality) to assess the design and generate data to allow design and fabrication of tooling (forming dies, molds, assembly fixtures, etc.).

Figure 21–5 shows an example of a full-size clay model for a planned production automobile. Built with great care, this model can be used as a master from which data can be scanned using a laser

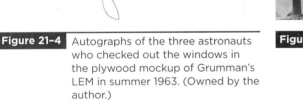

Figure 21-4 Autographs of the three astronauts who checked out the windows in the plywood mockup of Grumman's LEM in summer 1963. (Owned by the author.)

Figure 21-5 A clay model for the BMW 1 Series Convertible being worked on by the modeler. (*Source:* Produktcommunikation@bmwgroup .com; used with permission.)

coordinate measuring system to generate data for the tooling needed to produce it. Many automobile manufacturers produce a full-size plaster master. The master is not created until after automotive aerodynamics and crashworthiness testing has been completed on other models. Any final changes to the contours of the car are made on the master itself.

Masters are sometimes patterns, and sometimes patterns are masters. It is not uncommon in modern manufacturing for geometric data in the form of digitized coordinates to be taken from a full-size master via laser scanning or, less commonly, coordinate measuring machines using mechanical probes.

In Summary

Mock-ups are usually full-size models of a design that lack much—if any—functionality. They are intended to allow assessment of human factors; test customer reactions; serve as demonstrator, teaching, or training aids; or help in promotion and marketing. Their purpose is much more oriented to design than manufacturing. *Masters* are occasionally subscale but usually are full-size models used to aid in manufacturing by allowing design details to be added before tooling is designed from the master.

CHAPTER 22

Prototypes and Rapid Prototyping

IT IS NOT unusual that a word has different meanings for different people. The word *prototype* is such a word. Derived from the Greek *protos*, "first," and *typos*, "impression," come *prototypon*, "primitive form," and *prototypos*, "original, primitive," and thus one gets a sense of the word as it applies to engineering. A *prototype* is "an early sample or model [—that while it may be primitive—] is built to test a concept or process, or to act as a thing to be learned from and replicated."

In engineering, as in many other fields, there is inevitably some degree of uncertainty as to whether a particular new design will function as it is intended to function. Unexpected problems, unsuspected or unseen initially during the conceptualization stage of the design process, tend to emerge as the rule rather than as the exception. For this reason, a *prototype* is often used as an integral part of the product design process to allow designers and engineers to test theories and ideas, explore alternative designs, and, eventually, confirm performance prior to beginning production or construction (Figure 22–1). The prototype may then be refined—actually "tailored"—by engineers, largely based on their experience, to eliminate any errors, circumvent any deficiencies, and home in on desired functionality and performance. A common strategy involves an iterative series of trials and prototypes to test, evaluate, and then modify the design based on analysis of the earlier prototype.

Figure 22-1 A photograph of a collection of some of the prototypes submitted for what was to become the Jeep. The U.S. Army asked 135 companies to submit a prototype for a lightweight four-wheel-drive vehicle, allowing only 45 days for a response. Very few companies responded. The selected design was that of the Willys Overland Company. Following World War II, returning veterans clambered for surplus jeeps, so the Jeep went commercial, first with Willys, then with the American Motors Company, and then with Chrysler Motors, now part of Fiat. (*Source:* Patrick David Coovert, Mansville, Ohio, photographed at Fort Economy Sam Werner Museum Preservation Show in Hallsville, Ohio, with his kind permission.)

If a lot of this sounds familiar having read about "models" in previous chapters, that is because there is no general agreement as to what exactly constitutes a "prototype" versus a "model." In fact, the word *prototype* is often used interchangeably with the word *model,* and this can cause confusion. Different references, or industries, or companies within an industry, tend to divide prototypes into different categories, with some of these divisions following the division of models presented in this book. For example, one division involves five categories of prototypes, including:

1. *Proof-of-principle prototypes* (analogous to *proof-of-concept models,* discussed in Chapter 18).
2. *Form study prototypes* that allow designers to explore the basic size, look, and feel of a product without having to incorporate all—if any—of the eventual functions (analogous to *mock-ups,* discussed in Chapter 21).
3. *User experience prototypes,* of particular use in software design and development, but also used with hardware, allow assessment of the human-prototype interface (again, analogous to some *mock-ups*).
4. *Visual prototypes* intended to explore intended design aesthetics by simulating style, form, line, color, and surface texture (once again analogous to some *mock-ups*).
5. *Functional prototypes* (or *working prototypes*), which, as far as is practical, attempt to incorporate the final geometry, aesthetics, materials, and functionality of the intended design. In many cases, functional prototypes are most analogous to *test models* (Chapter 20), but may be unique models.

Prototypes can—and often do—differ from the ultimate production model in one or more of three fundamental ways:

■ *Materials:* To keep costs for the prototype low, especially since full functionality and performance may not be required, substitute materials are frequently used (e.g., plywood in mock-ups, ABS polymer in rapid prototypes)
■ *Processes:* To keep costs for the prototype low, and to circumvent the need for long lead time and costly tooling, prototypes often compromise by using more flexible processes (see "Rapid Prototyping," to follow).
■ *Lower fidelity:* To keep costs low for the prototype, nonessential details are often omitted from prototypes.

With ever-advancing computer-based modeling, it is becoming more practical to bypass the use of physical prototypes using virtual simulations—for better or worse. Computer modeling is now being used extensively in automotive design and in aerospace design. Boeing's 787 Dreamliner was developed almost entirely using computer modeling, with a full-size physical embodiment not made until production began. However, computer-based modeling has limitations. The delayed release of airplanes to customers can only make one wonder if problems were found in flight-testing of an actual 787 that were not brought to light by computer models.

Illustrative Example 22-1: Computer Modeling versus Physical Prototypes

When NASA approached Grumman Aerospace Corporation (and other contractors) in the mid-1970s for a design for a deployable, large space antenna that could fit in the cargo bay of the Space Shuttle in its folded configuration and be opened once in orbit, the company responded with one of its first computer-based (i.e., computer-aided) designs. State-of-the-art solid-modeling software of the time was used to design the antenna panels and folding linkages used to deploy the array. Animation of the model allowed deployment to be simulated.

Everything looked good, except to a couple of skeptical seasoned engineers who were

TABLE 22-1 Summary of the Various Types of Models and Prototypes Used in Engineering Problem-Solving

Modeling Technique	Main Purpose	Corresponding Prototype
Proof of concept	To demonstrate key principles using very simple models	Proof-of-principle prototype
Experimental model	To test new, unproven ideas or principles in designs or processes	[no counterpart]
Test model	To test a representative early or preproduction unit or system	Functional prototype (or working prototype)
Mock-up	To explore geometry (size/shape), and/or look and feel	Form prototype
	To explore aesthetics	Visual prototype
	To teach or train	Form or functional prototype
	For marketing	Form prototype

concerned about the lack of detailed kinematic analysis capability in the model to ensure that the linkages wouldn't lock up in the folded position. Since the antenna structure was being designed to minimize weight for use in a microgravity environment, the structure was incapable of supporting its own weight on Earth. Thus, making and testing a full-size model was not possible. A simple proof-of-concept prototype to reproduce the basic deployment mechanism convinced designers they needed to revise the design to prevent lockup at joints by incorporating a slight bias in the folded linkages for the opened array.

The design proposal was saved by a physical prototype, even as the age of computer-aided design (CAD) was emerging.

Table 22–1 summarizes the various modeling techniques used to solve problems, along with comparable prototyping techniques, where applicable.

Rapid Prototyping

Rapid prototyping involves the automatic construction of physical objects—often as the first physical realization of a considered design—using computer-aided additive manufacturing technology. When the first technologies became available in the late 1980s, only models or prototype parts could be created.[1] With advances in several of the various technologies available, production-quality parts can be produced as one-of-a-kind parts or in small numbers, so that rapid prototyping is now opening the field of *rapid manufacturing.*

The general approach in all rapid prototyping is to use *additive manufacturing technologies.* As the name suggests, these usually add material layer by layer as new 2D slices to build up a 3D shape.[2]

---

[1]Techniques for rapidly—and easily—creating 3D realizations from 2D slices are very old, and have been used, for example, by tool designers and builders for centuries. For example, tool-and-die makers frequently stacked thin sheets of metal cut to the shape of a 2D slice through a 3D object to create the 3D object after smoothing off corners. Wood boatbuilders used this technique for centuries in the creation of "loftings" (see Chapter 25). In fact, the Great Pyramids were built layer by layer, tier of stone by tier of stone, to produce the huge ediface. This may well have been what allowed artists to paint the walls of the burial chamber—without artificial light sources.

[2]There are newer techniques, such as *ballistic projection* that emulate ink-jet printing to create 3D shapes by spraying globules of molten or soft material.

Computer-aided design (CAD) data for a virtual 3D design is transformed into virtual horizontal cross sections or slices, and successive layers (or slices) are stacked until the full 3D prototype is complete. The added layers may be created from liquid, powder, or sheet materials, liquid materials having to harden or solidify, powder materials having to be consolidated, and sheet materials, perhaps, having to be bonded or otherwise joined.

The term *rapid* in *rapid prototyping* is relative, but is used to denote that the construction of the prototype can be done much more expediently than by using conventional methods to make a one-of-a-kind or small quantity of parts or articles.

A large—and growing—number of technologies are available for rapid prototyping. These are often competitive in the marketplace, but most have complementary capability so that several types may provide a large user much greater capability and flexibility. The various fundamental approaches (i.e., underlying techniques) used have been classified thus:

1. Techniques based on material addition
   1.1 Using a liquid
      1.1.1 Solidifying a liquid
         a. point by point
         b. layer by layer
         c. as a holographic surface
      1.1.2 Solidification [curing] of an electro-set or UV-set liquid
      1.1.3 Solidification of molten material
         a. point by point
         b. layer by layer
   1.2 Using discrete particles
      1.2.1 Joining powder particles by sintering
      1.2.2 Bonding powder particles with binders
   1.3 Using solid sheets of material
      1.3.1 Joining sheets by adhesive
      1.3.2 Joining of sheets by light, UV, or laser

Table 22–2 lists the various rapid prototyping methods along with the basic materials used and the additive processing method employed. Figure 22–2 shows two wonderful examples of rapid prototyping by two different methods. A great time-lapse video showing the 3D printing process (apparently using a laser-curing thermosetting polymer) can be seen on the Wikipedia.com site under the subject of "3D Printing."

Illustrative Example 22–2: Using Rapid Prototyping

A prospective client comes to the company where you work as an engineer involved in the design and manufacture of cast, as well as injection molded, metal and ceramic parts. He has a new design for the impellor for automobile water pumps that he shows you as a computer-based solid model. He asks your boss, the company owner, if he feels it can be produced by metal injection molding, and whether the company can produce it.

You ask if the CAD file for the solid model is available and whether you can download a copy onto your laptop. You ask to be excused from the meeting, and you return to your office. There you convert the CAD file to an STL file format used as the interface between CAD and the stereolithography rapid prototyping system you have in your prototyping shop.

Twenty minutes later you return to the meeting with an ABS plastic model spray-coated to look like metal. You show it to the client, who is obviously impressed. So impressed that

TABLE 22-2 Summary of the Most Common Rapid Prototyping Methods

Prototyping Technology	Base Material	Underlying Process
Laminated object manufacturing (LOM)	Paper	Thermal cutting + adhesive bonding (Classification 1.3.1)
Stereolithography (SLA)	Thermosetting polymer	White or UV light curing (Classification 1.1.2)
Fused deposition modeling (FDM)	Thermoplastics or eutectic metal alloys	Soften or melt followed by hardening or solidification (Classification 1.1.1 or 1.1.3)
Selective laser sintering (SLS)	Thermoplastic or metal powders	Solid-state diffusion bonding (Classification 1.2.1)
Direct metal laser sintering (DMLS)	Almost any metal alloy powder	Solid-state or transient-liquid phase sintering (Classification 1.2.1)
Electron beam melting (EBM)	Titanium alloy powders	Melting and solidification (Classification 1.1.3 a or b)
3D printing (3DP) (ballistic particle forming)	Various materials	Propulsion of molten or softened globules (Classification 1.1.3 a)

he asks to set up a meeting between his and your businesspeople to draw up a contract for an order for several thousand units.

Now it will be your job to use the 3D rapid prototype to develop molding dies for use in production.

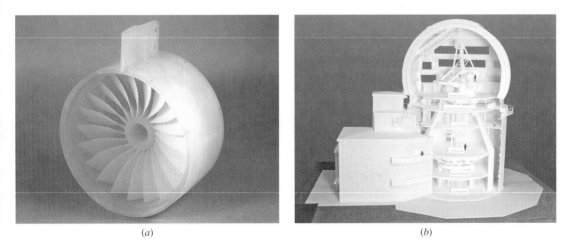

(a)   (b)

Figure 22-2 Photographs of two examples of rapid prototyping: (*a*) a model of the swirl tube's vaned disk produced from a design by NASA and MIT to create a new method for reducing the noise from commercial aircraft jet engines, made by stereolithography (SLA); (*b*) an early mock-up for the Advanced Technology Solar Telescope made by selective laser sintering (SLS). (*Source:* Both parts were created by Solid Concepts, Inc., in Townsend, Washington, for which photographs were provided and used with permission.)

In Summary

Prototypes mean different things to different people, since many different types are used for many different purposes. Most are physical models that represent a design, albeit perhaps not of the same material, or using the same processing, or with all of the nonessential details as the real item. *Rapid prototyping* is a modern means of automatically creating physical models quickly by using CAD data to generate a 3D model (or part) layer by layer as stacked 2D slices. Both prototypes and rapid prototyping are valuable physical techniques for solving certain engineering problems.

Suggested Resources

Books

Gibson, Ian, David W. Rosen, and Brent Stucker, *Additive Manufacturing Technologies: Rapid Prototyping and Directed Digital Manufacturing,* Springer Science + Business Media, Dordrecht, The Netherlands, 2009.

Nooran, Rafij, *Rapid Prototyping: Principles and Applications,* John Wiley & Sons, Hoboken, NJ, 2005.

Wright, Paul K., *21st Century Manufacturing,* Prentice-Hall Inc., Englewood Cliffs, NJ, 2001.

Website

Superb artistic examples of the use of rapid prototyping at www.georgehart.com/rp/rp.html.

CHAPTER 23

Trial and Error

THERE'S AN OLD adage that goes something like this: We sometimes learn more from our mistakes than we learn from our successes. True as this may be, it doesn't advocate failure, nor does it suggest that success without failure is not desirable and admirable. In the Discovery Channel's *Sons of Guns* reality series, three employees of Will Hayden's Red Jacket Firearms, one of the United States's most creative custom and specialty gun makers and sellers—among whom is Will's daughter, Stephanie— agree to decide which of them gets to do what all three want to do by seeing who can get closest to a small bull's-eye in a one-shot shoot-off. The young woman says, "You get to use the gun of your own choice." Each man enters the range with his own, personal-favorite handgun. When asked where her gun is, the young woman says she left it in her car and will get it after they each take their shot. They do, and both men are close, but neither hits the tiny bull's-eye. The young woman exits the shop's indoor range to retrieve her gun, She returns with a Russian Saiga 12-gauge tactical shotgun loaded with bird shot—upward of a hundred small metal balls packed in front of a heavy charge of gunpowder. The two men protest, but she reminds them, "No one said 'handgun.' That was your choice. This is mine!"

She fires and several pellets strike the bull's-eye, as most scatter to pepper the paper target with holes. She wins!

The moral, at least here, is: You're more likely to hit your target with a shotgun [approach] than with a single, carefully aimed shot [attempt]. But, a shotgun approach to problem-solving is risky and expensive and uncertain to succeed. Many more pellets miss (i.e., are errors) than hit (i.e., are successes).

Trial and error (or *trial by error*) is a basic method of problem-solving, for fixing things, or for obtaining knowledge. This approach contrasts with the other basic methods of problem-solving that use insight and theory. The former suggests brute force—and often elicits a shotgun approach. The latter suggests sophisticated preparation and contemplation to choose a well-aimed path. The former is quick and somewhat random. The latter is slow and deliberate. Between these two extremes lie many approaches to problem-solving like those discussed so far in this book. Known generically as *guided empiricism,* these approaches use theory to guide the method.

A *trial-and error-approach* to problem-solving can have merit, but—and it is a big *but*—trials should be guided by theory and thought to minimize errors.

A trial-and-error approach tends to be more successful with simple problems than with complex ones, and it is often the only choice when no apparent rules apply (as in some games). Use of trial and error need not—and should not—be careless. Rather, the user (for us, the engineer) can and should manipulate variables in an attempt to sort through possibilities that may lead to success.

There are several features of the trial-and-error approach to problem-solving that bear mentioning:

- The approach is *solution-oriented,* making no attempt to discover *why* a solution works, only that it *is* a solution.
- The approach is *problem-specific,* making no attempt to generalize a solution to other problems. (Hence, there can be little learning.)

- The approach is *usually nonoptimal,* attempting only to find *a* solution, not *all* solutions, and not the *best* solution.

It is these shortcomings that should discourage engineering faculty from teaching, advocating, or even allowing trial-and-error problem-solving, especially in design. Engineering students—especially early in their undergraduate education—have limited knowledge and very little to no experience in engineering. To make matters worse, increasingly, new faculty in engineering seldom have experience in engineering practice themselves. Thus, relying on trial and error teaches little but a certain lackadaisical attitude toward problem-solving and design. Worse, as suggested by the first three features of the approach just given, there is little learning. Trial and error is probably best left to seasoned engineers for two reasons: First, seasoning builds a strong knowledge base and a sense of what might and what might not work arising from experience-based intuition. Second, seasoned engineers are less likely to rely on trial and error in the first place, having learned many other approaches to problem-solving using *guided empiricism.*

Some of the shortcomings cited here may be overcome by keeping good notes and continuing to search for other solutions rather than stopping with the first solution. At least this way, there is some learning.

Trial and error is—or perhaps should be—a last-resort approach to problem-solving, as there are a number of problems with it. These include:

- It is tedious and monotonous.
- It is time-consuming—and, so, can be expensive and wasteful.
- There is an element of risk in that certain attempts at a solution could be very wrong and produce disastrous results. (If one doesn't know what one is doing, one can't predict the outcome!) (Figure 23–1).

Computers can help with trial-and-error problem-solving in that they are well suited to tedious, repetitive tasks. Furthermore, a bad solution on a computer that could be disastrous with physical trials might be okay.

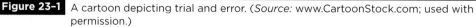

Figure 23-1   A cartoon depicting trial and error. (*Source:* www.CartoonStock.com; used with permission.)

Illustrative Example 23-1: Using Guided Empiricism

The coatings surrounding metal core wire in shielded metal-arc welding electrodes (commonly seen in use on construction sites for welding steel) have many functions, including:

- Providing a source of ions to allow easy arc initiation and stable action (from halogen or oxide-halogen salts with low ionization potentials)
- Generating nonoxidizing, air-excluding gases to protect both the core wire and the newly deposited weld metal from oxidation (from thermal decomposition of cellulose, limestone, or titania/rutile)
- Creating a reactive flux that helps refine the molten weld pool and, once solidified to form a glasslike slag cover, to protect the molten weld pool and still-hot solidified weld metal from oxidation (from glass-forming alkali or acid salts and other compounds, usually in the form of finely crushed minerals)
- Producing a slag cover that becomes solid before the weld pool solidifies to contain the molten weld metal when welding is done in vertical planes or overhead
- Addition of certain alloying elements to a simple carbon steel core wire, for example, to create an alloy steel weld deposit (from powdered alloying elements)

Designing—actually, formulating—such coatings is an extraordinarily challenging task, as it demands the designer or formulator to consider five, six, seven, or more input variables at the same time.

A common "trick" used by formulators, in lieu of random trial and error, involves using guided empiricism to evaluate the effect of various additions and combinations of ingredients one at a time or a few at a time—all at once. The approach involves using a very generic coating (known to work reasonably well on similar core wires from experience or from having reverse engineered similar electrodes [see Chapter 16] on the selected core wire. This "bare-bones" coated electrode is used to make a bead-on-plate weld along a plate on which has been deposited strip after strip of powdered ingredients being

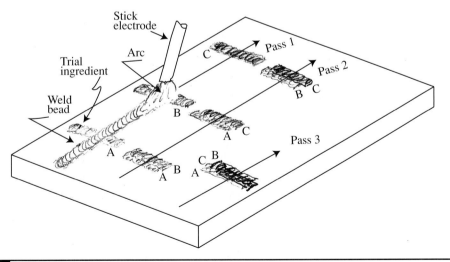

Figure 23-2 A technique used by formulators of the coatings for shielded-metal arc electrodes guides trial and error by making a bead-on-plate weld using a very simplistic coating on core wire cross strips of different prospective additional ingredients. The effect on the welding arc, weld pool, protective slag coating, and weld are assessed for each added single- or multiple-ingredient: A, B, C, A+B, A+C, B+C, and A+B+C.

considered for addition to the generic coating (Figure 23-2). As the arc—and weld pool—passes through each strip of ingredients that runs across the weld path, the formulator makes observations on the effects of each ingredient or combination of ingredients on arc and welding behavior, slag coverage and release (as slag must be removed from finished welds), and weld quality and weld metal properties. A second run may be made to try to refine the formulation to enhance certain observed effects.

Guided trials allow experienced formulators to achieve wanted results for a very complicated, multivariable problem using a minimum number of trials. The approach used here might also properly be considered a "model experiment" or "design experiment," provided a statistical method was used to select the trial ingredients and ingredient combinations to be evaluated.

In Summary

There is a place for *trial-and-error* approaches to problem-solving in engineering. A good guideline may be when there is little knowledge to be had—as opposed to information that is available but not known. Trial and error is a much safer approach when used by experienced engineers than by inexperienced engineers or engineering students. The reason is that experience helps guide trials and the motivation—or justification—for using this approach is not that it is easier, leading to a higher likelihood of a hit than a miss. Such trial and error uses what is known as *guided empiricism*.

PART THREE

Visual, Graphic, or Iconic Approaches to Problem-Solving

CHAPTER 24

Sketching and Rendering

"A PICTURE IS worth a thousand words." So goes an adage of uncertain origin but certain truth.[1] Many people respond to visual images more than they do to words, and a large fraction of people are what are called "visual learners," in that they think in and comprehend, absorb, and retain new knowledge from visual images better than from spoken (and heard) or written (and read) words (see Chapter 2).

Given this fact, it is not hard to understand that using visual techniques can be a valuable way to approach and solve problems. Such approaches are called *visual approaches* or, in scientific modeling terminology (see Chapter 2), *iconic approaches.*[2] These all use some form of physical visual aid or physical or mental visualization to seek solutions to problems—as well as to "see" the problem.

Clearly, engineers document and communicate designs for parts, devices, structures, assemblies, systems, and processes in the form of *engineering drawings,* a visual format. Once produced using mechanical instruments (from which came the term *mechanical drawing*), including T-squares, triangles, compasses, and French curves, to put top, side, and front orthogonal 2D views and/or isometric or perspective 3D views on vellum with pencil or ink, engineering drawings are now produced on computers using graphics software (AutoCAD, AutoDesk Inventor, Pro-Engineer, Solid Works, and others). While engineering graphics (EG) and computer-aided design (CAD) are essential skills for a modern engineer, they are *not* problem-solving techniques. Rather, they are communication techniques to allow designs to be executed (i.e., manufactured or constructed).

For the purposes of this book, this and the subsequent six chapters comprising Part 3 describe and discuss visual and graphic techniques for problem-solving. The first of these techniques is *sketching,* with brief reference to a close-related technique, *rendering.*

A *sketch,* in its simplest and most basic form apropos to problem-solving in engineering (and architecture), is a rapidly executed drawing that is not intended to be a finished graphic work, nor may it ever lead to any finished graphic work.[3] A sketch may have any of several purposes, including:

- To record something one sees that is, or could be, important to recall later
- To quickly capture a thought or idea (that may flash into—and out of—one's mind) on paper in the form of a very simple image (perhaps to be refined or made more detailed later)
- To graphically demonstrate or communicate an idea, a principle, or an image to another

In engineering and science, there are two universal "languages" by which one engineer or scientist communicates to another, irrespective of each one's mother tongue: mathematics (covered

---

[1] Modern use is attributed to Fred R. Barnard's ad "One Look Is Worth a Thousand Words" that appeared in the December 8, 1921, issue of *Printer's Ink.* He used a similar ad in the March 10, 1927, issue, "One Picture Is Worth Ten Thousand Words," and attributed the saying to an ancient Chinese proverb, which was not true, but he thought it would help people take it more seriously. The adage became inaccurately associated with Confucius.

[2] A key, if not *the* key, to the early success of Apple and its Macintosh personal computer was the use of *icons.* The rationale was to make the PC more user-friendly by making interaction with it more natural and less abstract than had been the case with the prevailing approach driven by IBM.

[3] The origin of the word *sketch* is the Greek *schedios,* which means "done extemporaneously." From this definition, it is clear that sketches are likely to be hand-drawn.

(a)   (b)

Figure 24-1 (a) The concept sketch for the Sydney Opera House in Sydney, New South Wales, Australia, by the Danish architect, Jørn Utzon, dated 01 January 1956, and (b) a photograph, from the same perspective. Note how Utzon captured the essence of the design in his simple—probably extemporaneous—hand sketch. (*Sources:* Mitchell Library, State Library of New South Wales at library@sl.nsw.gov.au; used with permission of the library.)

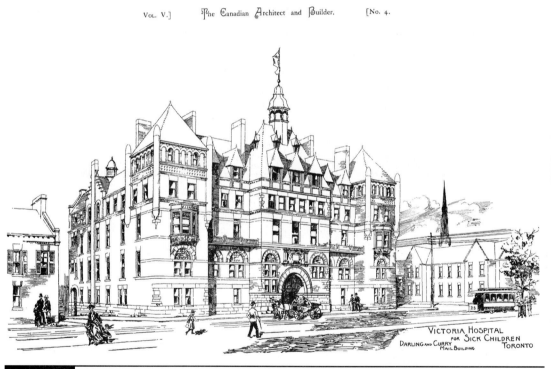

Figure 24-2 A finely detailed architectural sketch of Victoria Hospital for Sick Children, Toronto, Ontario, Canada (*Source:* From *The Canadian Architect and Builder,* Volume 5 [1892], Issue 4, Plates 3a and 3b, in the U.S. public domain because it was published before January 1, 1923; in Wikimedia Commons, originally contributed by Skeezix1000 on 5 December 2008 and placed in the public domain, so free to use.)

in Part 1) and physical visualizations (covered in Parts 2 and 3). *Sketching* is the simplest visual means of communicating because it is meant to be done quickly and simply. For this reason, having—or developing—an ability to sketch is a key skill for an engineering student and engineer. Most formal courses in engineering graphics include some lessons on and exercises in sketching, for example, to teach scaling, perspective, and so on. If there is a program in architecture at one's university, it is not at all unusual to see first-year architecture students sitting on benches or on a grassy spot on campus and sketching one of the buildings. Depending on the instructor's lesson, the sketch might be very simple (Figure 24–1, *left*) or very elaborate (Figure 24–2). Simple sketches are intended to capture the essence of a design, while elaborate sketches are intended to capture details or, perhaps, evoke feelings. Whether for young architects or young engineers, both aspects of—and techniques of—sketching are important. The former aspect and technique are to quickly capture or express an idea or an object by omitting anything that detracts from the key impression to be had or expressed. The latter are to train one's eyes (and brain) to see all there is to see and to develop a degree of discipline that, hopefully, extends to one's thinking process.

As a technique for problem-solving, sketching should be done mostly for one's self—to capture what might end up a fleeting thought or idea. Sketches should appear in every engineer's notebook, ranging from very simplistic, like that by Danish architect Jørn Oberg Utzon (April 9, 1918—November 29, 2008) of the concept for what was to become the easily recognizable Sydney Opera House at the harbor in Sydney, New South Wales, Australia, in Figure 24–1, to slightly more detailed, like that of the first

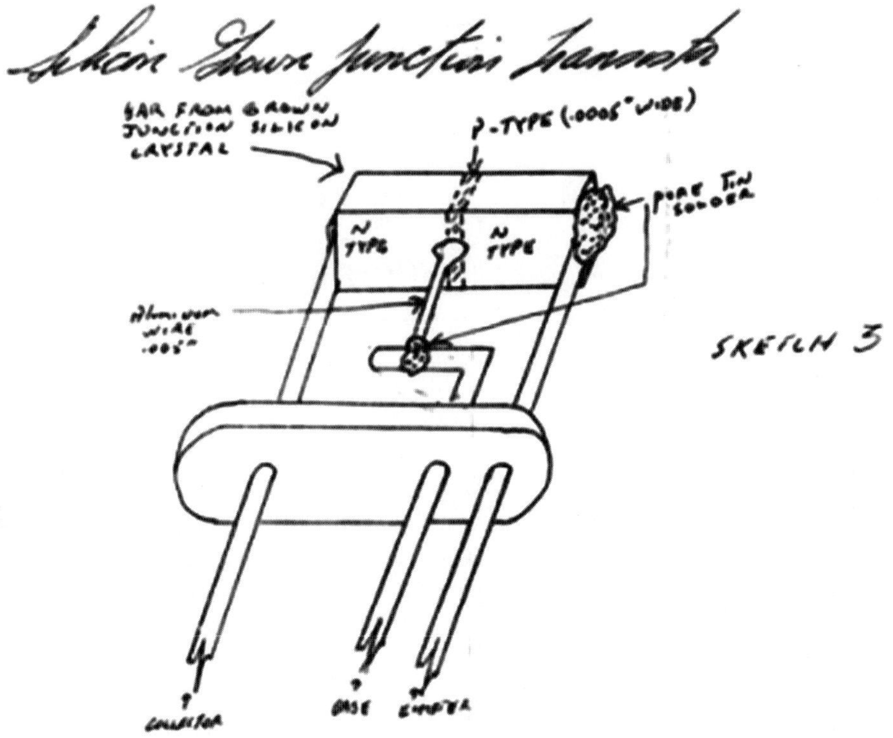

Figure 24-3 A sketch made by Texas Instrument engineer Mort Jones in his notebook on May 6, 1954, showing the concept for what was to become the first grown-silicon transistor. This simple sketch was a harbinger of the revolution that was to sweep the world of electronics in the future. (*Source:* Texas Instruments, used with the company's permission.)

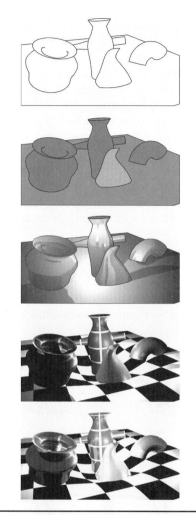

Figure 24-4 A variety of rendering techniques applied to a single 3D scene. (*Source:* www.wikipedia
.com, attributed to Maximillian Schönherr, 8 May 2005; used under the GNU Free
Document License version 1.2.)

grown silicon transistor by Texas Instrument engineer Mort Jones in his notebook on May 6, 1954, that
harbingered the revolution to come in the world of electronics, in Figure 24–3.

With modern computers pervasive in schools, residences, and workplaces, one might be tempted
to use any of the several software packages that allow sketching. But sketching, by its etymology, is
supposed to be extemporaneous, and, as a skill for engineering, it should be in the engineer not in a
computer!

Rendering has come to be known as a process of generating an image from a model (or models in
a scene file) by means of computer programs (Figure 24–4). This technique for creating more detailed
and, potentially, highly realistic images is valuable and can have a role in problem-solving, at later
stages, when details and realism may be needed or useful (see Figure 20–6). But one should know that
many seasoned engineers are superb renderers, even when hand-drawing an object or a scene—perhaps
to a fault. Such people draw what they see and not, as artists do, what they feel.

The interested reader is encouraged to seek out references on computer-based rendering.

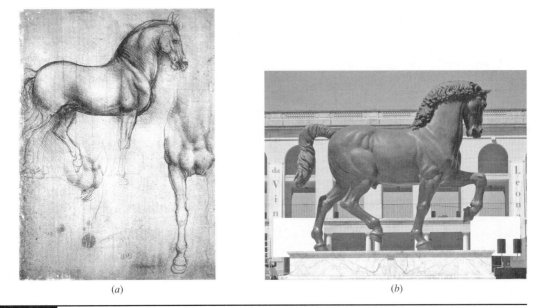

(a)   (b)

Figure 24-5 (a) Leonardo da Vinci's sketch for *Il Cavallo,* a planned 24-foot-high, 70-ton bronze cast sculpture that was never made by the artist, and (b) a photograph of a replica by Nina Akamu and the Tallix Art Foundry, Beacon, New York, in Milan. (*Sources:* Wikimedia Commons; the sketch was originally contributed by Kasper2006 on 4 March 2009 under Creative Commons Attribution Share-Alike 2.0 "Generico"; the photograph was originally contributed by Sturmbringer on 20 March 2005 and was released by the author to the public domain. Both are used under free license.)

Illustrative Problem 24–1: Sketching in Problem-Solving

The master of sketching for problem-solving was, without question, the genius Leonardo di ser Piero da Vinci (April 15, 1452–May 2, 1519). The archetype of the Renaissance man,[4] he was an accomplished painter, sculptor, musician, writer, architect, mathematician, scientist, engineer, inventor, anatomist, geologist, botanist, and cartographer. Leonardo used sketching not only as an artistic medium, in and of itself, but as a way of capturing the innumerable ideas that flourished in his imagination. His notebooks are filled with sketches of inventions (e.g., machine gun, giant crossbow, airplane, rotorcraft, tank, submarine), some built and tested, many not, and studies for paintings (e.g., *The Last Supper* and the *Mona Lisa*) and sculptures (e.g., *Il Cavallo,* the great unfinished bronze horse; Figure 24-5a).

Leonardo worked for 17 years (from 1482) to create a 24-foot-high 70-ton bronze cast sculpture of a horse for the Duke of Milan, Ludivico of Moro, to honor his father, Francesco. Despite numerous sketches for the sculpture itself, to get the anatomy correct and the pose dramatic, and for the steel framework needed to support a heavy, full-size clay model (or prototype, per Chapter 23), Leonardo never finished the planned bronze sculpture because of technical difficulties he worked on with his left-brain analytical genius and his right-brain artistic genius.

---

[4]The Renaissance spanned the period from the fourteenth to the seventeenth century, getting a major kick with the invention of the movable-type printing press by Johannes Gensfeisch zur Laden zum Gutenberg (circa 1398–February 3, 1468) in 1439—as it launched the first "information age" by making printed books more widely available (growing from fewer than 2 million before the invention to 20 million in Europe alone by 1500)—and may have reached its zenith with da Vinci.

Finally, in 1997, after more than two decades of efforts to complete the unfinished work, begun by Charles Dent (until his early death from Lou Gehrig's disease), picked up by Garth Herrick part-time in 1988, and finally completed, with major modifications from the Dent-Herrick design, by animal artist Nina Akamu, hired by the Tallix Art Foundry of Beacon, New York, a great bronze statue was completed. It now stands in Milan as a tribute to Leonardo (Figure 24–5*b*).

Amazingly, a fairly recent study using computer-based software for modeling the casting process proved that Leonardo's design could have been produced in a single pour of 70 tons of bronze, exactly as he had calculated.

In Summary

Having an ability—whether natural or developed with effort—in *sketching is invaluable for engineers to solve problems. It allows ideas to be captured quickly and to be shared extemporaneously, when needed. Rendering* a highly realistic image can have value for some purposes, although it is usually outside the skill set and responsibility area for engineers.

CHAPTER 25

Tracings and Transfers, Templates, Lofting and Lines-Taking

THERE ARE TIMES when, for a variety of reasons, it is necessary to move information in the form of points and/or lines either from a physical object to a drawing or vice versa. Modern computers have made this task easier for many situations and some subset of users. Two-dimensional images can be scanned to convert an image, pixel by pixel, into a digital data file that can later, at will, be reproduced as a facsimile image. Three-dimensional objects can also be scanned using a laser or other structured-light source or even touch-probe to digitize points, often as lines across the surface, and converting these into a data file or using digital data in a computer file to create a 3D object. The latter is the basis for computer-aided design/computer-aided manufacturing (CAD/CAM).

However, not every situation lends itself to capturing point, line, and surface information this way *and*, believe it or not, not every business has or wants to have a computer capability to perform this task. For such situations—and businesses, craftspeople, or engineers—there are, fortunately, a number of tried-and-true techniques for accomplishing this task on the way to solving a problem, often in the creation of a design. Many of these techniques, not surprisingly, have been around for a long time.[1] A list of the most important techniques includes:

- Tracings
- Transfers
- Templates
- Lofting
- Lines-taking

The list may not be complete, but it is surely representative of these important, related techniques for solving engineering problems using a graphical approach and/or methodology.

Let's look at each in turn.

Tracings and *transfers* are related in that they are both ways of moving graphical information from a 2D image or object or vice versa, respectively. A *tracing* is a graphic copy of an image or object made on a translucent sheet (e.g., *tracing paper*) for the purpose of recording that image for its own value (e.g., historical or archival value) *or* for using the tracing to create a reproduction of the original. The actual operation may use a pencil, pen, marker, crayon, chalk, or powdered charcoal to copy the image or object onto the tracing paper by overdrawing the lines of the original or, for objects with relief, rubbing

---

[1]There is evidence that the ancient Egyptian architect for the Great Pyramid at Giza used 3D scale models from which line data were taken and scaled up using dividers, for example.

Figure 25-1 Photograph of Michelangelo's masterpiece, *The Creation*, painted on the ceiling of the Sistine Chapel from 1508–1512 using both tracings and transfers. At the very center (here, just left of the center), the figure of Adam can be seen reaching up to God as he is infused with life by God's touch. (*Source:* Wikipedia, submitted by Aaron Logan in June 2005; used under the Creative Common Attribution License, 2.5.)

over the original. A *transfer* moves a design or original image to a new surface for the purpose of reproducing that design or original. The actual operation uses an easily transferable charcoal or colored wax drawing of the original that is then transferred to the new surface by rubbing.

Illustrative Example 25-1: Using Tracings and Transfers

Michelangelo di Ludovico Buonarroti Simoni (March 6, 1475—February 18, 1564), commonly known simply as Michelangelo, was a gifted Renaissance painter, sculptor, architect, and engineer from Florence (Firenza), Italy.[2] His white Carrara marble sculptures of the Pieta and of David, made before Michelangelo was thirty, are considered two of the most magnificent sculptures ever created for their soft, lifelike quality.

In 1505, he was invited to Rome by the newly elected Pope Julius II and, in 1508, was commissioned to paint the ceiling of the Sistine Chapel off St. Peter's Cathedral in Vatican City. His technique involved making tracings from his full-size drawings laid out on tables on the floor of the chapel, making pinholes along every line, and after placing the pin-pricked tracing over the right area of the complexly vaulted ceiling, tamping a bag filled with crushed charcoal on the tracing to transfer black charcoal dots to the pre-prepared plaster-overlaid ceiling. These he used to guide his painting of the fresco masterpiece (Figure 25-1).

His true genius is reflected in his ability to distort the shapes and proportions of his full-size 2D drawings to compensate for the distance the viewer would be from the complexly contoured 3D ceiling so that, when viewed from the floor of the chapel, the images all appear in proper proportions.

---

[2]Michelangelo designed and led the construction of the great dome of St. Peter's in a masterwork of architecture and engineering. Check out a color image of this masterpiece on the Internet, and visit the Sistine Chapel to be awed by Michelangelo's genius.

Both tracings and transfers have been—and continue to be—used by skilled craftspeople and engineers in areas such as tool and die making. Some people interested in genealogy or history, or just having a morbid fascination with cemeteries, reproduce the faces of inscriptions on tombstones using what are called *rubbings.* A dark crayon or charcoal stick rubbed on a sheet of paper placed over the face of the tombstone creates dark areas where the paper contacts the tombstone and leaves white areas where letters or images are recessed into the tombstone face.

A *template,* while once exclusively referring to what are now known as *mechanical templates,* are aids used to represent a finished article for one or more purposes during that article's creation. Examples include:

- A stencil, pattern, or overlay used in graphic arts (e.g., sign making, drawing, painting) and sewing to replicate shapes, designs, or letters, often made of paper (if temporary) or of more durable material (if permanent)
- A predeveloped page layout in electronic or paper media used to make new pages with a similar design, pattern, or style
- form letter, as a predefined letter which retains its primary intent when individually customized or personalized with variable data or text

and, as intended here, for engineers:

- A mechanical template of paper, cardboard, wood, plastic, or sheet metal cut and, perhaps, formed to the shape of a product to be created or checked (e.g., as for a die, mold, or formed part)

Templates may be flat patterns from which either a flat or shaped or contoured three-dimensional part is made by copying the template in a more durable structural material. To create a 3D object, a flat pattern template may be used to create a flat, 2D replica that is then folded and/or formed, or multiple flat pattern templates may be pieced together to create the 3D article, say, by welding sheet-metal pieces together. Such flat patterns are commonly used in the fabrication of sheet-metal heating, ventilation, and/or air-conditioning ducts. Some readers may be familiar with the use of stiff paper templates to help create compound-curved fuel tanks or fenders from cut, formed, and welded pieces of sheet-gauge steel for motorcycles on Discovery Channel's *American Chopper.*

Besides being used to aid in the fabrication of an object by being copied and, therefore, replicated in a more durable or expensive material, templates can be used strictly to check the geometric and dimensional accuracy of a fabricated object. These are sometimes known as *check templates* and are commonly used to check the proper achievement of complex, compound-curved surfaces, as found in dies and molds and even on manufactured articles such as boat hulls, aircraft fuselages or, especially, wings, or automobiles.

An example of the use of a clear plastic template to check the contour of a finished railroad car is shown in Figure 25–2.

The use of *lofting* and *lines-taking* has had—and still has—particular utility in wood boat-building. Perhaps not so surprising, lofting was also used in the early days of airplane building, particularly for amphibious aircraft. In both boat hulls and aircraft fuselages, compound-curved shapes are used for low drag, but with the penalty of difficult fabrication. Again, these two techniques, like tracings and transfers, involve reverse processes. *Lofting* is a drafting technique used to lay out curved lines for the fabrication of curved pieces and/or construction of compound-curved shapes, such as hulls. The full process during wood boat-building, as an example, goes like this[3]:

1. First, the hull lines are reproduced full size on the lofting floor of the shop using the designer's scale drawing and a table of offsets from reference lines along the centerline of the hull, above and below a waterline, and fore and aft from some reference plane near the bow (Figure 25–3).

[3]The kindness of master wood boatbuilder and designer Paul Gartside, of Paul Gartside Ltd., in beautiful and historic Shelburne, Nova Scotia, is gratefully acknowledged.

Figure 25-2 Photograph showing workers at Siemens using a large, clear-plastic template to check the contour of a newly fabricated body for a passenger train car for accuracy against the design drawing. (*Source:* Siemens Corporation; used with the courtesy and permission of Siemens Press Picture.)

2. Once the hull lines have been faired (i.e., blended to produce a smooth, continuous line with correct shape for the boatbuilder's trained eye) and checked, full-size patterns (for use as templates) are picked up off the floor (Figure 25–4). Temporary building molds are also made directly on the loft floor.

Figure 25-3 Dimensions of key points on a scale drawing by the designer are transferred to the floor of a shop, or *loft,* using appropriate scaling factors (see Chapter 10). These points are measured from reference or datum lines in the three orthogonal directions of the hull. (*Source:* Paul Gartside Ltd., Shelburne, Nova Scotia, with the kind permission of master boatbuilder and designer Paul Gartside.)

Figure 25-4 The boatbuilder blends, or "fairs," lines to produce smooth curves that are streamlined and aesthetically pleasing. Temporary building molds are made right on top of the *lofting.* (*Source:* Paul Gartside Ltd., Shelburne, Nova Scotia, with the kind permission of master boatbuilder and designer Paul Gartside.)

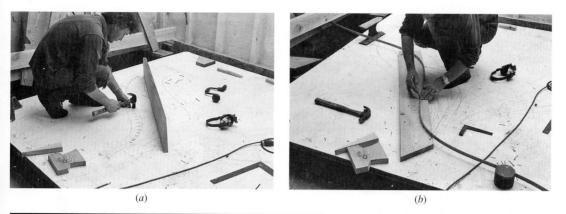

(a)   (b)

Figure 25-5 A pair of photographs showing how the shape of lines on the lofting are lifted onto material from which parts of the boat or parts of the jig for building the boat are made. The technique uses nails with their heads laid along the lines (*a*) and then creates an imprint of the nail heads demarcating the lines on wood by heavy blows from a hammer (*b*). (*Source:* Paul Gartside Ltd., Shelburne, Nova Scotia, with the kind permission of master boatbuilder and designer Paul Gartside.)

3. The standard technique for picking up the shape of the lines from the floor involves laying nails on their sides (at appropriate spacing to capture the line shape) with their heads on the line and tapping them into the floor (Figure 25–5*a*). The piece of mold material is then laid on top and given a few heavy blows. The mold piece is then turned over and nails are driven in the impressions left by the nail heads. A batten is then sprung around these nails and a pencil line is drawn (Figure 25–5*b*). The piece is then cut out on a band saw. The process may sound complicated, but with practice, boatbuilders find it to be quick and accurate.
4. The pieces of building mold are then assembled on the drawing (Figure 25–6).
5. When the molds have been made, they are set up to form the jig on which the boat is built (Figure 25–7*a* and *b*).

For modern fiberglass hulls, the same basic techniques are used to create the mold.

The reason for the continued use of this age-old graphic technique is that computers blend, or *fair,* digitized points together using mathematically based splines (see Chapter 5), which are generally not as pleasing to the eye as fairing done by the eye of a classic designer or, especially, a stylist.

Figure 25-6 Photograph showing pieces of the building mold for the hull being assembled on the lofting. (*Source:* Paul Gartside Ltd., Shelburne, Nova Scotia, with the kind permission of master boat builder and designer Paul Gartside.)

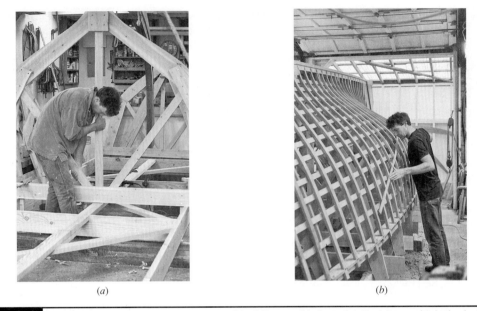

(a)   (b)

Figure 25-7 Photographs showing the erection of building molds into the jig (a) on which the hull will be created by applying shaped boards or planks (b). (*Source:* Paul Gartside Ltd., Shelburne, Nova Scotia, with the kind permission of master boat builder and designer Paul Gartside.)

Lines-taking is the reverse process: taking detailed measurements of an existing boat's shape to create a table of offsets (from a baseline). Lines-taking can be used to reproduce an exact copy of the original or a scaled version—up or down in size. Figure 25–8a shows the design for a wood boat, along with, in Figure 25–8b, the table of dimensions to be transferred to the lofting floor.

In manufacturing, especially in the creation of tooling (e.g., fixtures, dies, or molds), mechanical *templates* are used. These are full-size layouts, on cardboard, wood, or sheet metal, used to make reproductions. Templates were widely used in building the airframes for airplanes. Templates are still used for this purpose as well as to check contours of finished production items for accurate manufacturing.

In Summary

Tracings and transfers and lofting and lines-taking are related techniques for converting 2D images, objects, or design to 3D objects or designs, and vice versa. *Tracings* make copies of 2D images or objects through overlaid paper, while *transfers* move an image onto a 2D or 3D surface by rubbing an easily transferable flexible medium onto the new surface. *Templates*, specifically *mechanical templates*, are specially made aids used to create or check a shape, especially when the shape of the part or full object is complicated in geometry. *Lofting* creates a full-size layout for a curved or compound-curved 3D shape using a gridwork of measurements offset from a baseline, while *lines-taking* collects offset measurements from a baseline on a 3D shape to create plan, profile, and cross-sectional views. All are invaluable techniques for solving certain kinds of problems in engineering involving both design and manufacturing.

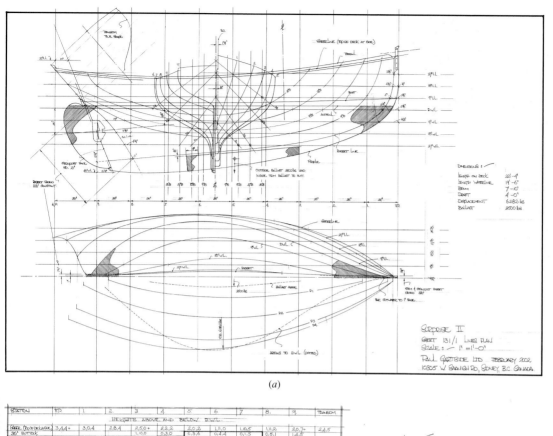

(a)

STATION	FP	1	2	3	4	5	6	7	8	9	TRANSOM
HEIGHTS ABOVE AND BELOW DWL											
SHEER (TOP OF SHEERLINE)	3,4,4+	3,0,4	2,8,4	2,5,0+	2,2,2	2,0,2	1,11,0	1,10,5	1,11,2	2,0,7+	2,4,5
36" BUTTOCK				1,0,0	0,3,0		0,3,6	0,4,4	0,5,1	0,5,1	1,4,5
27" BUTTOCK			2,1,2	0,2,6	0,6,6	0,10,1+	0,0,0	0,6,7	0,0,6	0,8,0+	1,6,0
18" BUTTOCK			0,5,2	0,8,0	1,1,5	1,3,5	1,2,7	0,11,6	0,5,2	0,4,1	0,10,4
9" BUTTOCK		0,0,4	0,8,6	1,4,4	1,8,5	1,9,7	1,8,7	1,5,3+	0,10,5	0,0,7+	0,6,3+
RABBET LINE		0,9,2+	1,7,4	2,1,0	2,4,4	2,7,0	2,9,2	2,11,4	2,10,4	0,2,5	0,2,0
PROFILE		1,8,0	2,1,7	2,9,3	3,3,0	3,7,1	3,10,1	3,11,7	4,0,0		
HALF BREADTHS											
SHEERLINE		1,4,2+	2,5,2	3,1,7+	3,7,3	3,10,2	3,10,7	3,9,5	3,6,4	3,1,5	2,5,2
27" LL			1,1,7	2,3,4	3,1,3						
18" LL		0,11,2	2,0,1	2,10,3+	3,5,5+	3,9,5	3,11,0	3,9,6+	3,6,4	3,0,4	2,3,0
9" LL		0,8,2+	1,7,7	2,6,4	3,2,7	3,8,1	3,10,0	3,8,5	3,3,4	2,4,0	1,2,6
DWL		0,5,2	1,2,5	2,0,7	2,9,6	3,3,5	3,5,3	3,2,2	2,4,5	0,6,3	
9" WL			0,8,6	1,4,6	2,0,2	2,4,6	2,4,7	1,11,1	0,11,4		
18" WL			0,2,5	0,7,4	1,0,0	1,2,0	1,0,5	0,8,3+	0,3,4		
27" WL					0,4,1	0,5,2	0,4,7+	0,3,5+	0,1,7		
RABBET LINE		0,1,6	0,1,6+	0,2,4+	0,3,4	0,4,1	0,3,6+	0,2,7	0,1,6	0,1,6	0,1,6
DIAGONALS											
D1			0,7,3+	0,11,6	1,2,5	1,3,5	1,2,7	1,0,3	0,7,4+		
D2		0,9,3+	1,6,2+	2,0,6	2,4,6	2,6,3	2,5,6+	2,3,0	1,9,0	0,11,1	0,6,2
D3		1,3,2+	2,3,3	3,0,5+	3,7,2+	3,10,6	3,11,1	3,8,4	3,2,7	2,6,2	2,0,1
D4		1,4,7+	2,6,1	3,4,0	3,11,6	4,4,6	4,6,5	4,5,0	3,11,6+	3,4,1+	2,8,3

NOTE:

OFFSETS ARE GIVEN IN FEET, INCHES, EIGHTHS
TO THE OUTSIDE OF PLANKING.

LOFT FULL SIZE.

SURPRISE II

SHEET 131/2 TABLE OF OFFSETS
PAUL GARTSIDE LTD   FEBRUARY 2002
10805 W SAANICH RD SIDNEY, B.C. CANADA.

(b)

Figure 25-8 A scale drawing of the hull for a custom wood boat (*a*), along with the table of offsets used to transfer the scale drawing to the lofting floor in full size (*b*). (*Source:* Paul Gartside Ltd., Shelburne, Nova Scotia, with the kind permission of master boatbuilder and designer Paul Gartside.)

CHAPTER 26

Graphing and Graphical Methods

PART 3 OF this book is entitled "Visual, Graphic, or Iconic Approaches to Problem-Solving." It could equally appropriately have been entitled "Graphical Methods for Problem-Solving." It wasn't, because of semantics, which involves subtle but important differences in the interpretation of words by some people as opposed to others.[1] The intent of Part 3 is to present, describe, discuss, and exemplify approaches (as techniques) for solving problems that use physical *images*—as opposed to physical three-dimensional objects, as in Part 2—either to take advantage of the way many people think (i.e., using visualization and imagery) *or* to help focus and direct thinking by first "seeing" the problem and then using visual approaches to find viable solutions.

A little reflection on the titles of chapters within Part 3 should reveal that *all* of the techniques presented involve imagery and/or visualization, while some (but not all) have a basis in mathematics (approaches that are covered in Part 1). For example, *sketching* (the principal subject of Chapter 24) is purely a technique to allow visualization, with no basis in mathematics, while *rendering* (also covered in Chapter 24) also allows visualization but has a strong basis in mathematics (e.g., ray-tracing and boolean algebra) when computer-based, as in CAD. Likewise, *lofting* and *lines-taking* (subjects of Chapter 25) are techniques to aid in visualization, each of which has a direct basis in solid and descriptive geometry of the object (e.g., boat hull).

In this chapter, the techniques of *graphing* and *graphical methods* have a mathematical basis and are intended to allow visualization. *Graphing* will be familiar from prior experience in secondary school, as well as from Chapter 3 (under "Equation Forms") and from Chapter 12 ("Response Curves and Surfaces"), while *graphical methods* may not be familiar. In the context of this chapter, *graphical methods* are restricted to those techniques that can be used to solve problems either more easily and naturally than by using more abstract mathematics *or* for which a mathematical approach is impossible or intractable, even though there is a mathematical basis for what is involved.[2]

With this explanation of what to expect, let's proceed.

Graphing

A *graph* in mathematics is a plot of the values for a function f as the collection of ordered pairs of values—or, on a set of coordinate axes, coordinates—for example, $(x, f(x))$ in two dimensions. For a pair of cartesian coordinate axes, in which the horizontal x axis is typically the independent variable and the vertical $f(x)$ (or y) axis is typically the dependent variable, the function plots as a curve on the cartesian

---

[1] In the case of the title for Part 3, as elsewhere in this book, the author attempted to use language that is clear and meaningful to the reader while still being accurate within the discipline of engineering and within the subject matter involved.

[2] *Graphical methods* and *graphical solutions* are terms that are also used for advanced mathematical approaches for solving problems involving optimization via linear programming. This chapter does *not* address this context of graphical methods.

TABLE 26-1 Major Graphing Formats or Graph Types and Their Particular Value	
Format or Type	**Particular Value**
Scatter plot	To see the presence or absence of a pattern in data
Line plot	To see the form of a function and/or the trend in the data
Bar chart	Typically as a histogram, to show the statistical distribution of data or to easily display comparisons
Pie chart	To easily see and compare the relative contribution or importance of factors

2D plane. For three orthogonal cartesian axes, in which x and y axes are both independent variables that form a horizontal plane and the $f(x,y)$ or z axis is vertical and represents the dependent variable, the function plots as a 3D surface. In science and engineering (as in other fields, such as business), *graphing* is a technique (and graphs are tools) used for many purposes.

Graphing as briefly covered in Chapter 3 helps make apparent the form of the equation, which can be invaluable in problem-solving by depicting behavior in a readily perceived and understandable fashion and/or for relating seemingly unrelated phenomena or process. Graphs also reveal the form of the equation to allow the underlying physics (or causative or operative mechanism) to be revealed. In Chapter 12, graphing is shown to be valuable in problem-solving by vividly showing the sensitivity of the output of a process (for example) to control parameters using response curves or surfaces. Response curves or surfaces allow easy identification of ranges for input (or control) parameters that give relative stability to the output.

Since graphing in problem-solving in engineering and science is intended to provide a simple display of behavior, engineering students and engineers need to recognize how different plotting formats can and should be used to provide the most meaningful display and powerful impact. Basic formats are presented in Table 26–1, along with a particular advantage of each.

Illustrative Example 26–1: Using Graphing

Simplistic analysis of the behavior of a material (in a structure) subjected to cyclic loading (i.e., fatigue) tends to consider only what stress can be tolerated for a required number of cycles of life, as S-N plots. A serious shortcoming of this approach is that it presumes (or assumes) the material/structure is initially free of any defects (stress raisers from geometric features or from surface cracks or internal flaws such as shrinkage cracks in welds or castings, nonmetallic inclusions in metal and alloy product forms processes by melt metallurgy, gas pores in castings, etc.), which it seldom if ever is.

A more realistic approach is to presume a flaw *is* present that is just smaller than the largest flaw that can be reliably detected during nondestructive examination by x-radiography, for example (see Illustrative Example 7-3). This approach considers the fracture toughness behavior of the material,[3] in which inherently tougher materials are able to tolerate larger flaws for a known prevailing stress intensity or, said another way, a structure can tolerate a higher stress intensity for the smallest flaw that can be reliably detected. The data

---

[3] *Fracture toughness* is the property that indicates how well a material resists the sudden onset of unstable propagation of a surface or internal physical flaw, at the speed of sound in the material.

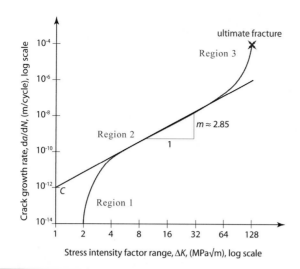

Figure 26-1 A generic plot of fatigue crack growth rate d*a*/d*N* versus stress intensity range Δ*K*. (*Source:* Wikimedia Commons; originally contributed by Tomeasy on 23 July 2010, used under Creative Commons Attribution Share Alike 3.0 Unported free license.)

used for life prediction in fatigue environments is crack growth rate per cycle (d*a*/d*N*) as a function of stress-intensity factor range (as Δ*K*).

Figure 26–1 shows a typical d*a*/d*N* versus Δ*K* plot. The generic curve shows a distinct pattern of behavior. In fact, for lower values of Δ*K* (Region 1), slow crack growth is exhibited, asymptotically approaching zero growth rate as the stress-intensity range approaches zero. For a significant midrange of Δ*K* (Region 2), d*a*/d*N* exhibits linear power-law behavior, and, therefore, being linear, is quite predictable. At high values of Δ*K* (Region 3), rapid unstable crack growth prevails, asymptotically approaching infinite growth rate as the stress-intensity range moves upward.

The lesson here is: The graphical display of the data reveals regions of different behavior, the middle one of which is very predictable within realistic stress-intensity factor ranges for practical designs.

Graphical Methods

It is rare in engineering that it is not helpful to *visualize* the essence of a problem in the consideration of an approach and technique(s) to reach a solution. In fact, for some problems, a pure mathematical approach might be impossible, intractable, or just overly and unnecessarily complicated.[4] Using a *graphical method* of solution may then be a viable—if not the only—option. *Graphical methods* most often involve 2D representations of what may be 3D situations or phenomena, which are used directly to generate a solution from the graph or graphic or indirectly to aid in any actual mathematics. This being said, a few examples (in the form of illustrative examples) are given that represent different areas for the use of a graphical (versus analytical) approach. The list is not complete but should make clear the idea. Covered examples include:

- Free-body diagrams (for depicting forces and resolving vectors, based on vector algebra or vector calculus)
- Venn diagrams (for showing relationships among sets and subsets of information, based in the mathematics of logic and set theory)

[4]Recall the author's admonition: Never measure a manure pile with a micrometer. In engineering, good enough is often what it is all about.

- Mohr's circle (for depicting and analyzing the state of stress at a point in a material, based on tensors)
- Ewald's sphere (for depicting allowable diffractions in crystals, also based on vectors)
- Load-deflection diagrams (for depicting the behavior of springs, based on geometry and algebra)

Illustrative Example 26-2: Using Free-Body Diagrams

A *free-body diagram* or *force diagram* is a graphic depiction used by physicists and engineers to analyze the applied and reaction forces and/or moments acting on a body of interest. Creating a free-body diagram always makes it easier to understand the forces and moments in relation to one another and either suggests how to proceed mathematically or, for some problems, allows a solution to be extracted directly from a scaled diagram.

Forces are vectors, having both a magnitude and a direction. Hence, the free-body diagram aids in resolving those vectors—and associated forces—into horizontal and vertical (e.g., shear and normal) components and into a resultant vector (force).

Imagine a block of mass m on a ramp with an incline angle of θ degrees (Figure 26-2). Gravity acts on the mass to produce a vertical downward force $F_g = mg$. This force resolves into normal and shear components on the face of the ramp, with the normal force N being the reaction to the gravitational force created by the ramp against the block, a shear component down the ramp that would cause the block to slide if the ramp-block interface was frictionless, and, if not, a frictional shear force F_f that acts in the opposite direction along the ramp surface to resist sliding. The normal force $N = -F_g \cos \theta = -mg \cos \theta$, acting opposite F_g (taken to be positive). The frictional force $F_f = -\mu N$, where μ is the coefficient of friction at the ramp-block interface, which acts in the opposite direction of the shear component of mg acting down the ramp incline (i.e., $+mg \sin \theta$). If friction is present, the block will slide or stay put depending on the relative values of μ and θ. The solution comes from comparing $mg \sin \theta$ to $\mu mg \cos \theta$, with the block staying put when the frictional force equals or exceeds the force acting down the ramp face due to gravity, in other words, for $\mu \geq \sin \theta / \cos \theta$.

Free-body diagrams are *always* valuable in solving problems involving forces and moments, even (or especially) if mathematics is used to calculate the results.

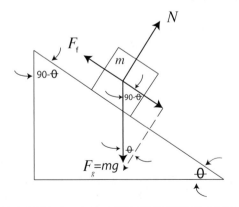

Figure 26-2 A typical free-body diagram, here, for a block of mass m on an inclined ramp. The gravitational force $F_g = mg$ is shown, along with the reaction normal force N between the block and the ramp's surface, and the friction force F_f attempting to keep the block in place against the force of gravity. Resolution of the gravitational force into components normal to and along the incline are also shown.

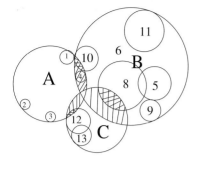

A. Engineering metal and alloys
    1. Intermetallic compounds/long-range-ordered alloys
    2. Metal glasses/amorphous metals
    3. Foamed metals and alloys
B. Engineering ceramics
    4. Metalloids and semiconductors
    5. Fired porous ceramics (e.g., bricks and tiles)
    6. Cement and concrete
    7. Steel-reinforced concrete (intersection of 6 with A)
    8. Glasses
    9. Glass ceramics (crystallized glasses)
    10. Carbonaceous materials (e.g., graphite)
    11. High-performance ceramics (Al_2O_3, SiC, Si_3N_4, etc.)
C. Engineered Polymers (polycarbonate, nylon 6-6, A-B-S, etc.)
    12. Elastomers (e.g., natural and synthetic rubbers)
    13. Foamed polymers (e.g., Styrofoam)

Figure 26-3 A graphical representation of the fundamental types and subtypes of materials important to engineering, in the form of a Venn diagram. The sizes of the circles and overlaps are proportional to the usage of the various types and subtypes in the year 2010. The legend identifies the various material types and subtypes

Illustrative Example 26-3: Using a Venn Diagram

A *Venn diagram* or *set diagram* is a 2D graphical representation that shows all possible logical relationships or solutions (as sets) between a finite aggregate of things. Venn diagrams[5] are used in mathematics to teach set theory within the area of logic, and involve boolean algebra. They are frequently used in engineering and other disciplines to show the relationship between or among various sets and subsets of information.

Figure 26–3 portrays the relationship between the fundamental types and subtypes of materials, according to the author in his textbook *The Essence of Materials for Engineers* (Ref. Messler).

The three fundamental types (i.e., *metals, ceramics,* and *polymers*) are shown by circles of a size proportional to the amount of each material used (in 2010). Within the fundamental types, as sets, are subsets of other materials (also proportional in size to the approximate amount of the particular material used) that are not fundamentally different from the set to which they belong. Circles (i.e., sets or subsets) that overlap represent either materials that exhibit properties that are shared by or between two fundamental types or physical/mechanical (as opposed to chemical) combinations of the fundamental materials represented by each circle.

Venn diagrams can help in solving problems by logically grouping and organizing information pertinent to the problem.

Illustrative Example 26-4: Using Mohr's Circle

Christian Otto Mohr (1835–1918), a German civil engineer, proposed the use of a two-dimensional graphical representation of the state of stress at a point in a material/structure. In *Mohr's circle,* the horizontal axis (abscissa) represents the normal stress σ, while the vertical axis (ordinate) represents the shear stress τ. The circumference of the circle is the locus of points that represent the state of stress (as combinations of normal and shear stresses) on individual planes of any orientation. Mohr's graphical technique was logically

[5] Venn diagrams were introduced in 1880 by John Venn (1834–1923), a British logician and philosopher.

extended to 3D stress states in which the three principal stresses characterizing the stress state are given on the abscissa as σ_1, σ_2, and σ_3 (Figure 26–4).

Of particular interest in mechanics are what are known as *principal stresses* and *principal directions.* For these directions, there are no shear stresses, only normal stresses (i.e., the principal stresses are invariant). The normal stresses equal the principal stresses when the stress element is aligned with the principal direction, and the shear stress equals the maximum shear stress when the stress element is rotated 45° away from the principal direction.

A two-dimensional Mohr's circle can be constructed if one knows the normal stresses along the *x* and *y* axes (i.e., σ_x and σ_y,) and the shear stress (τ_{xy}) lying between these axes, these stresses forming a 2×2 tensor. Values of key features found on Mohr's circle can be found mathematically (as opposed to strictly graphically using a scaled Mohr's circle) by the following relationships:

$$\sigma_1 = \sigma_{max} = 1/2(\sigma_x + \sigma_y) + \{[1/2((\sigma_x - \sigma_y)]^2\}^{1/2} + \tau_{xy}{}^2$$

and

$$\sigma_2 = \sigma_{min} = 1/2(\sigma_x + \sigma_y) - \{[1/2(\sigma_x - \sigma_y)]^2\}^{1/2} + \tau_{xy}{}^2$$
$$\sigma_{ave} = (\sigma_x + \sigma_y)/2$$
$$R = \tau_{max} = \{[(\sigma_x + \sigma_y)/2]^2 + \tau_{xy}{}^2\}^{1/2}$$

The maximum and minimum principal stresses (σ_1 and σ_2, respectively) and the average stress (σ_{ave}) can be found on the horizontal axis, and the maximum shear stress (R or τ_{max}) at the bottom of the circle.

There is a 3D version of Mohr's circle, for which the details of the mathematics are left to the interested reader to find online, if desired.

Mohr's circles are invaluable in mechanics and have particular utility in what is known as *yield criteria.*

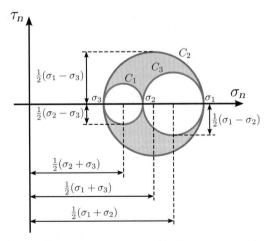

Figure 26-4 Mohr's circle for a three-dimensional stress state at a point in a material/structure. Principal stresses characterizing the stress state are identified on the abscissa as σ_1, σ_2, and σ_3. (*Source:* Wikimedia Commons; originally contributed by Sanpaz on 15 December 2008, used license free, as the author released the image to the public domain.)

Illustrative Example 26-5: Using Ewald's Sphere

The German-born U.S. physicist and crystallographer Paul Peter Ewald (1888–1985) conceived of a geometric construct used in the diffraction of x rays, electrons, or neutrons by crystalline materials. *Ewald's sphere* graphically demonstrates the relationship between:

- The vector for the incident and diffracted beams of radiation
- The diffraction angle for a given reflection or diffraction event
- The reciprocal lattice of the crystal (in which the spacing between atoms—actually, planes of atoms—along the three axes of the crystal system are plotted as points separated by the reciprocals of these distances)

Ewald's sphere is used for 3D space, while Ewald's circle is used for 2D simplifications.

The theory underlying Ewald's construct is left to the interested reader to find for himself or herself. Suffice it to say here, any point in the reciprocal lattice that lies on the surface of a circle or sphere of radius $2\pi/\lambda$ gives rise to a diffraction event (and peak). (See Figure 26–5.)

Illustrative Example 26-6: Using Load-Deflection Diagrams

Problems involving springs or material structures acting with a "spring constant" k (the proportionality constant in Hooke's law for force versus extension for a spring, $F = kx$) can be difficult to impossible to solve purely analytically. The usual approach is to employ force-deflection diagrams. A *force-deflection diagram* is a graphical depiction, as a plot, of the deflection of a spring or springlike material or structure as a function of the applied force. Examples of situations for which a force-deflection diagram has particular value as a problem-solving technique or analysis aid are (1) actual behavior of a spring or, especially, as a system of springs or a compound (multielement) spring and (2) a bolted joint, in which a threaded bolt is stretched elastically (by application of torque) so that a stack-up of sheets or plates is elastically compressed to prevent unwanted separation. In the case of a bolted joint, the bolt and the joint elements have different "spring factors" or spring constants, so each responds (i.e., stretches or compresses) differently. In the case of a compound spring, each individual spring may have a different spring constant and, as a consequence, stretches or compresses differently. The total response of the system of

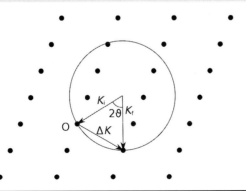

Figure 26-5 Ewald's sphere, in a generic depiction. Planes (shown as points in a reciprocal lattice) that give rise to a diffraction event for an x ray (for example) of wavelength λ (giving rise to a vector K' of magnitude $2\pi/\lambda$) lie on the surface of the sphere (here, shown only as a circle). (*Source:* Wikimedia Commons; originally contributed by wiso on 3 February 2005, used under Creative Commons Attribution Share Alike 3.0 Unported free license.)

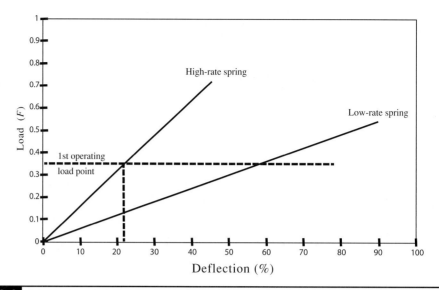

Figure 26-6 A typical load-deflection diagram for two springs with different spring constants—or stiffnesses—one high and one low. For simplification here, a particular "Operating load" (shown by the horizontal dashed line) will cause different deflections in each spring, shown by projecting the intersections of the "Operating load" line with the sloping spring-rate lines to the Deflection axis. Similarly, for a particular deflection (shown by the vertical dashed line), each spring supports a different load, shown by projecting the intersections of the deflection line with the sloping spring-rate lines. Readers should check the Internet for a detailed explanation of analysis using load-deflection plots for springs with different rates or for tension-loaded bolts.

springs or a hybrid spring requires the use of a force-deflection diagram to allow graphic analysis of the overall behavior.

Figure 26-6 shows the force-deflection diagrams for two springs of different stiffness or spring constant. Imagine how much simpler it would be to use a force-deflection diagram for springs that act in series, more stiff ones only after less stiff ones extend (or compress) to a certain extent, than using mathematics. A diagram in which the early behavior is plotted for low loads and, then, the later behavior is plotted as the load increases allows the total deflection to be easily picked off the compound plot.

In Summary

There are many problems in engineering for which a graphical representation clarifies the problem and/ or makes easier the finding of a solution. These techniques may involve *graphing* (as plots of functions or data) or *graphical methods* that, as used here, have a basis in mathematics but employ geometric constructs to aid visualization and solution.

Suggested Resources

Beer, Ferdinand P., and K. Russell Johnson, Jr., *Mechanics of Materials,* McGraw-Hill, New York, 1992.

Messler, Robert W., Jr., *The Essence of Materials for Engineers,* Jones & Bartlett Learning, Sudbury, MA, 2010.

CHAPTER 27

Layouts

THE DICTIONARY DEFINES the noun *layout* as "an arrangement or a plan, especially the schematic arrangement of parts or areas." Implicit in the definition—and in a layout itself—is a purpose, which is often to achieve a logical ordering or flow.

In engineering, *layouts* may be an end goal for or an output of an effort *or* may be used to simply organize thoughts and thinking on the way to solving a problem. Either way, for either purpose, layouts are important in engineering, as well as in architecture.

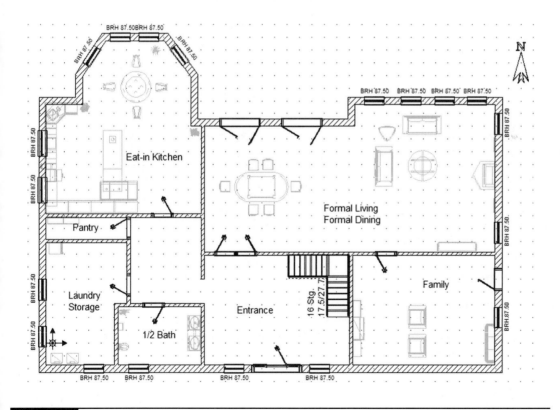

Figure 27-1 An architect's schematic layout or plan for a single-family home. (*Source:* Wikimedia Commons; originally contributed by Boereck on 21 March 2006 and placed in the public domain for free use.)

Some of the objectives of layouts in engineering include:

- The optimum utilization of space (e.g., land or floor area) or material in a sheet form
- The necessary and appropriate distribution of energy (e.g., electricity, gas, and oil) or services (e.g., water; fresh, cooled, or heated air; cable TV; fiber-optic communication)

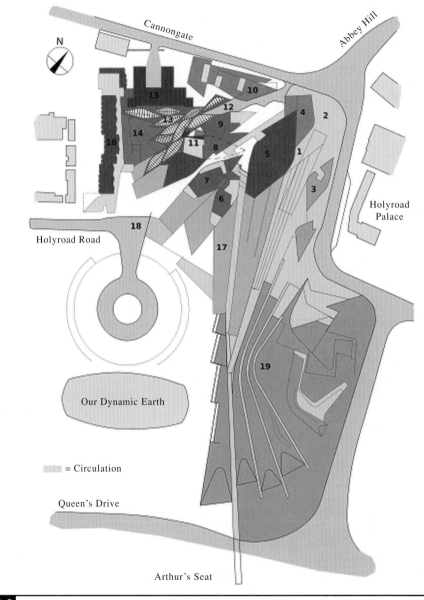

Figure 27-2 An artist's site plan (or layout) for the Scottish Parliament: (1) public entrance, (2) plaza, (3) pond, (4) press tower, (5) debating chamber, (6) tower one, (7) tower two, (8) tower three, (9) tower four, (10) tower five Cannongate Building, (11) main staircase, (12) MSP's entrance, (13) lobby, (14) garden, (15) Queensbury House, (16) MSP building, (17) turf roof, (18) car park and vehicular entrance, (19) landscaped park. (*Source:* Wikimedia Commons; originally contributed by Paul McGinn and converted to SVG by DTR on 22 September 2007, and placed in the public domain to allow free use.)

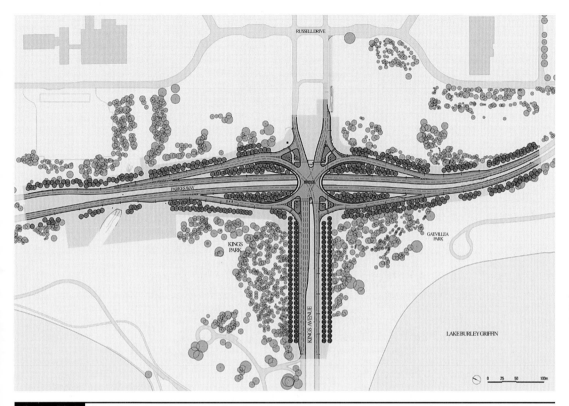

Figure 27-3 A civil engineering layout for the Kings Avenue Intersection in Canberra, the capital city of Australia in the Australian Capital Territory. (*Source:* Image supplied courtesy of the National Capital Authority. Commonwealth Copyright. All rights reserved.)

- The logical and efficient flow of vehicles (e.g., traffic on highways, trains on railways), people (e.g., walkways, pedestrian overpasses, people-movers, escalators), sewage, material and parts (e.g., in a manufacturing facility), and so on
- The arrangement of operations or items in consideration of human factors and ergonomics (e.g., the layout of a computer keyboard)
- The strategy or strategies for achieving some goal (e.g., a layout for a battle plan in military science)

Layouts are used by most engineering disciplines as well as by architects. Examples include:

In architecture:
- To arrange building sites or lots on land and/or of buildings or homes on sites or lots for efficient utilization, convenience, aesthetics, and so on (Figures 27–1 and 27–2)
- To divide working or living space for need, efficiency, economy, aesthetics, and so on.
- To arrange walkways (and possibly roadways)
- To arrange landscape (e.g., plantings, walls, fountains, waterfalls, etc.)

In civil engineering:
- To arrange roads and entrance and exit ramps (Figure 27–3)
- To arrange sewers (to allow gravity flow) and utility services (electricity, water, gas, etc.)
- To arrange buildings and other structures in an integrated industrial complex
- To arrange some types of industrial plants (with inputs from experts in the plant's operation)

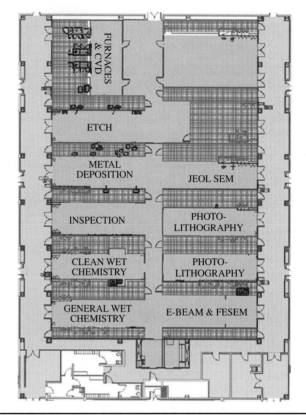

Figure 27-4 A layout of the NanoFab Cleanroom Floor at the National Institute of Standards & Technology (NIST) at its Center for Nanoscale Science & Technology in Gaithersburg, Maryland, USA. (*Source:* Found online via a Google image search of "floor plans" and used freely under the public domain for government material.)

In industrial engineering:
- To arrange floor space for orderly and logical movement of people and flow of raw materials and product (Figure 27–4)
- To arrange automated manufacturing and assembly lines (e.g., transfer lines)

In electrical engineering:
- To distribute electrical power service (e.g., distribution grids, municipal power)
- To arrange wiring (Figure 27–5)
- To arrange circuit paths and components on microelectronics printed circuit boards for optimum functionality and to maximally utilize space

In mechanical engineering:
- To distribute ventilation and cooled (i.e., conditioned) and heated air
- To arrange machinery logically in a plant

In chemical engineering:
- To arrange flow of raw materials, chemicals, and finished product in a processing plant (for which a superb example showing the use of 3D solid modeling or rendering can be seen on YouTube.com under "3D As-Built Model Chemical Plant")

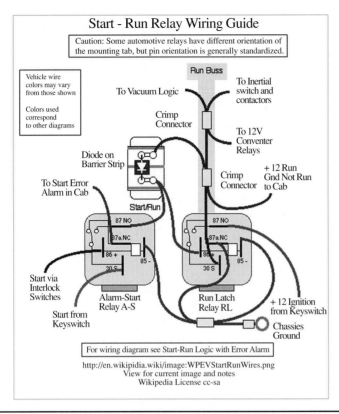

Figure 27-5 A typical wiring layout for a modern automobile "start-run" relay. (*Source:* Wikimedia Commons; originally contributed by "File Upload Bot" [Magnus Manske] on 18 January 2010, under Creative Commons Attribution Share-Alike 1.0 free license.)

There are sometimes fine lines between *layouts* and *designs* and *plans*. As used here, layouts tend to be schematic or illustrative, while designs and plans tend to be carefully dimensioned to allow actual construction. As but one example, precisely where a flowering dogwood tree is planted in a landscape layout is seldom critical, but where windows or doors are placed for proper natural lighting or entrance/egress may be.

Illustrative Example 27–1: Using Layouts

When you plan to install an in-ground sprinkling system for watering the lawn around your home, you need to create a layout. There are two major considerations to be dealt with, one obvious and one less obvious but equally important.

First, you obviously want the sprinkler heads to provide water coverage for the entire lawn. Any area that doesn't receive water will dry out, turn brown, and, eventually, die. Thus, you need to create a layout that allows you to place the sprinkler heads so that you get some (but not excessive) overlap, not missed areas, and, ideally, no more heads than are necessary (Figure 27–6). You can do this by laying out circles representative of the watered area for a head on a scaled drawing of your property (e.g., a simple scaled sketch on quad-ruled graph paper). Fortunately, heads can be adjusted for full-, half-, or quarter-circle operation.

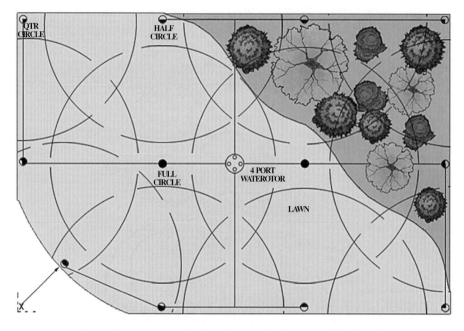

DIAGRAM D - BASIC LAYOUT - FULL, PART AND QUARTER SPRINKLERS

Area covered by sprinklers at maximun spadings is show by broken lines.
Overlap of sprinklers is necessary for proper coverage. Failure to overlap will result in dry spots.
Placing part circle sprinklers on edges and throwing inwards reduces overspray on paths and walls.

Figure 27-6 | Simple schematic layout for an in-ground sprinkler system to ensure complete coverage without wasted water spraying on walkways, and so on. (*Source:* Lawler Specialty Products Pty. Ltd., Marleston, South Australia, at waterotor.com; used with permission.)

Second, you need to consider water flow. This means you'll need to divide the arrangement of sprinkler heads you came up with into logical groups, each of which is fed water by a separate line from a manifold back near the water source. This is to minimize a drop in water pressure along a line when water is delivered to each head in the line—and to further minimize pressure drops created with every angle or bend in a line.

While not "rocket science," this is a problem with multiple constraints and an optimum solution. Hence, short of turning a do-it-yourself project into a major engineering project, a simple layout that offers flexibility for making changes on the fly would be best.

In Summary

Layouts are schematics to show the general arrangement of parts or areas. They may be an end goal of an effort by an engineer, but they are usually used on the way to a final design, plan, or problem solution to guide thoughts and thinking.

CHAPTER 28

Flow Diagrams and Flowcharts

THREE TYPES OF DIAGRAMS are used by engineers (as graphical methods) to aid in the understanding of problems or processes involving many variables or operational steps. These three types are:

- Flow diagrams or flowcharts
- Causal diagrams or cause-and-effect diagrams
- Decision diagrams or decision trees

Each has a distinct purpose, so each is presented and discussed, in order, in the last three chapters of Part 3.

A *flow diagram* or, more commonly, a *flowchart,* is a type of 2D graphical display that represents a process or an algorithm,[1] showing the steps as either boxes of various shapes, symbols, or icons, with their

---

[1] An *algorithm* is defined as "a step-by-step problem-solving procedure." In mathematics, it is further defined as "an established, recursive computational procedure for solving a problem in a finite number of steps."

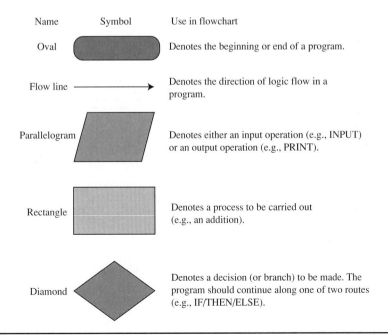

Name	Symbol	Use in flowchart
Oval		Denotes the beginning or end of a program.
Flow line	→	Denotes the direction of logic flow in a program.
Parallelogram		Denotes either an input operation (e.g., INPUT) or an output operation (e.g., PRINT).
Rectangle		Denotes a process to be carried out (e.g., an addition).
Diamond		Denotes a decision (or branch) to be made. The program should continue along one of two routes (e.g., IF/THEN/ELSE).

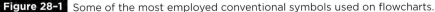

Figure 28-1 Some of the most employed conventional symbols used on flowcharts.

199

order shown by connecting arrows. The value of flowcharts is that they can assist greatly in the step-by-step solution of a given problem, including troubleshooting a multistep process that goes awry. They are especially useful in understanding, designing, documenting, or displaying a complex process or algorithm involving many steps.

For processes, the boxes, symbols, or icons are generally used to show the individual operations or activities. Icons may appeal to visual learners, while boxes tend to appeal more to analytical minds. There is an established convention for the shape of the box, symbol, or icon to use for different things (Figure 28–1). For example, rectangular boxes are used to depict individual process operations. Diamond-shaped symbols are used to depict decision points. The arrows that connect the boxes, symbols, or icons represent the flow of material, parts, product, and control information and/or data.

A simple flowchart for checking why a faulty lamp is not operating is shown in Figure 28–2.

For algorithms, the symbols represent steps in the solution to a problem, often in the form of a question for which a "Yes" or "No" response leads (by an arrow) to the next appropriate step. An example for calculating the factorial of N, N!, is shown in Figure 28–3.

Like other types of graphical displays, flowcharts help the engineer visualize what is going on. This aids in understanding the overall process—or an overall solution approach employing a mathematical algorithm—and can help in finding erroneous, missing, or unnecessary steps. For manufacturing processes, flowcharts can also help find choke points or bottlenecks to production throughput.

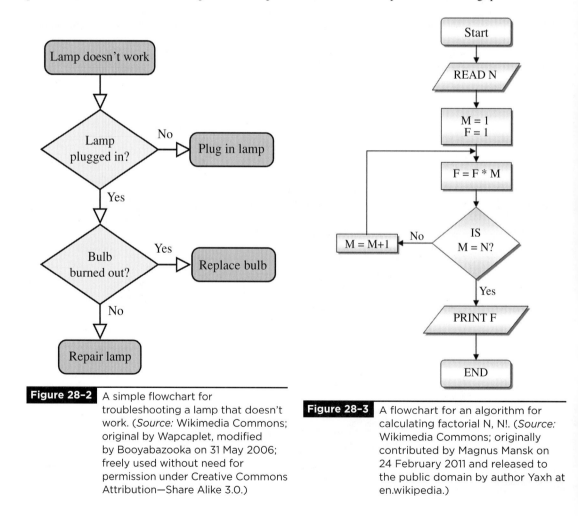

Figure 28-2 A simple flowchart for troubleshooting a lamp that doesn't work. (*Source:* Wikimedia Commons; original by Wapcaplet, modified by Booyabazooka on 31 May 2006; freely used without need for permission under Creative Commons Attribution—Share Alike 3.0.)

Figure 28-3 A flowchart for an algorithm for calculating factorial N, N!. (*Source:* Wikimedia Commons; originally contributed by Magnus Mansk on 24 February 2011 and released to the public domain by author Yaxh at en.wikipedia.)

Computers and modern commercially available software greatly facilitate the construction of a flowchart, but, as a word of caution, thinking through the process flow beforehand is always a good idea.

Besides their use by engineers in problem-solving or troubleshooting, flowcharts are commonly used in the management of processes. Four general types of flowcharts for management of processes suggested by Allen B. Sterneckert in *Critical Incident Management* are:

1. *Document flowcharts* to show controls of document flow through a process or associated with a process
2. *Data flowcharts* to show controls over data flows in a system
3. *System flowcharts* to show controls at a physical or resource level
4. *Program flowcharts* to show controls in a program with a system

These flowcharts are more management tools than problem-solving techniques, so are not of direct interest here. They are valuable to know about, however, for engineers in practice.

Figure 28–4 shows a flowchart for the process of manufacturing tires using easily recognizable icons. Figure 28–5 shows a flowchart for an information management help desk.

Illustrative Example 28-1: Using a Flowchart

Batches (consisting of hundreds) of automobile security lug nuts that employ a machined multilobed groove on the top face of the heat-treated steel nut to prevent removal without a matching "lock key" wrench were failing final inspection due to cracking at the ends of lobes. Curiously, all batches of defective nuts were found to have had the security grooves

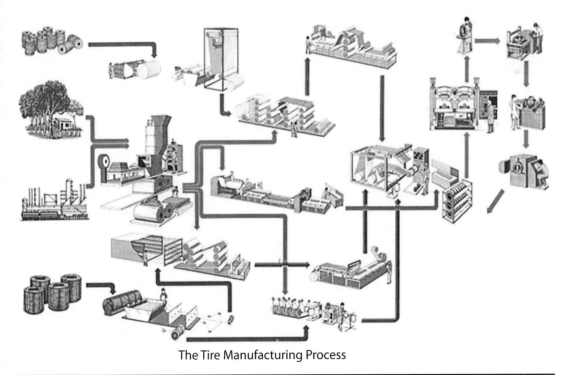

The Tire Manufacturing Process

Figure 28-4 A flowchart for the manufacture of tires. (*Source:* www.buytyres.com; used with permission.)

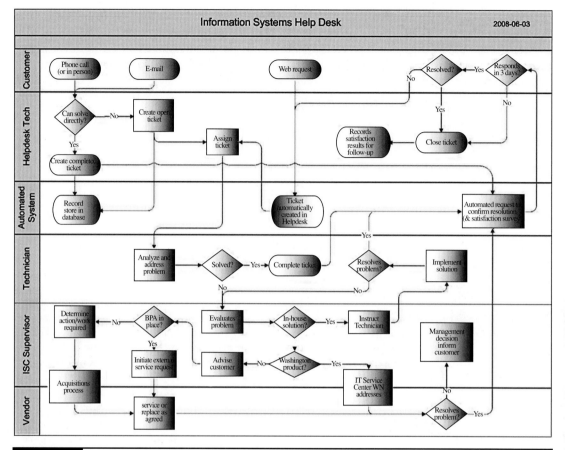

Figure 28–5 | A flowchart for an information management help desk (*Source:* Originally contributed by author MPRI Sandra on 23 December 2008, and released by the author to the public domain at en.wikipedia.)

machined on the same milling machine, one of three identical machines used in production. Cracking was detected during a scheduled inspection step following electroplating with decorative hard chromium.

Metallographic examination of nuts containing cracks revealed traces of hard chromium down inside many of the bigger (i.e., wider) cracks, indicating that the cracks were present either before plating or were produced early in the plating process. In addition, following etching with 2% nitric acid in methanol (2% nital), all cracked nuts exhibited martensite—an extremely hard phase of steel formed following heating into the so-called austenite region (at around 870°C [1600°F]) followed by very rapid cooling (say, by quenching into water)—surrounding the cracks, while nuts free of cracks exhibited the normal strong, tough pearlitic structure that was supposed to exist in the bar stock from which the nuts were fabricated. A review of the flowchart for the manufacture of the security lug nuts showed that hard chromium plating immediately followed the milling of the multilobed grooves. This suggested that "quench cracks" were formed when martensite was produced due to excessive frictional heating and rapid cooling during machining of the multilobed grooves *or* "hydrogen-assisted cracking" was occurring in as-quenched martensite during the electroplating process. Either way, the cause was as-quenched martensite from severe frictional heating—but only for one (of three) machines.

Interviews with the operator of the machine on which defective nuts were being produced revealed that he had been verbally directed by a supervisor to try to extend the life of milling cutters past the normal change-out frequency previously used and still used on the other two machines. This resulted in wear in these cutters that produced more frictional heat during machining—so much heat, in fact, that at the end of every loop in the multilobe groove, martensite was being formed.

A flowchart was a tremendous aid in tracing the point in the multiple-step manufacturing process where cracking originated. Root-cause determination (e.g., as-quenched martensite cracking) took some additional detective work to find out why only one of three identical milling machines was producing as-quenched martensite.

In Summary

Flowcharts are graphical displays of multiple steps or process operations in manufacturing, for example, or steps for algorithms used in solving a mathematical problem, often using a computer program. Some flowcharts use boxes and some use icons, but all are arranged to show the flow of material, activities, and/or actions using arrows. Flowcharts for algorithms typically include decision, in other words, "Yes" or "No," paths.

Suggested Resources

Sterneckert, Alan B., *Critical Incident Management,* Auerbach Publications, Boca Raton, FL, 2003.

CHAPTER 29

Causal Diagrams or Cause-and-Effect Diagrams

JUST AS IT is often important or useful for engineers to identify and order the steps and flow of material, product, or information while trying to solve a problem or troubleshoot an errant process involving multiple steps, it is also often important or useful to know cause-and-effect relationships between variables or factors in a problem or a process. The former need is fulfilled by *flowcharts,* while the latter need is fulfilled by *causal diagrams.* Both are visual or graphic techniques for problem-solving.

A *causal diagram* is a method—often used as a tool—to allow the visualization of cause-and-effect relationships between variables in a real process or system or in a model of a process or system. Each variable, step, or operation in a process or system is depicted by a *node.* Any variable that has a causal relationship with another variable is denoted by an arrow pointing toward the variable(s) on which it has a causal effect.

There are two general types of causal diagrams:

- Causal loop diagrams
- Fishbone or Ishikawa diagrams

Causal loop diagrams (CLDs) tend to be used more in business and commerce than in engineering, although they are used in engineering and in science. They aid in visualizing how interrelated variables, factors, or events affect one another using a set of nodes (one for each variable, factor, or event) and arrows to show the direction of effect(s). Arrows (and causal effects) are labeled as positive (+) or negative (–) depending on whether the effect is reinforcing (+) or not (–). A positive causal link indicates the two connected nodes change in the same direction. That is, if one increases, the other increases, and vice versa. A negative causal link indicates the two connected nodes change in opposite directions. That is, if one increases, the other decreases, and vice versa.

Figure 29–1 shows a typical causal loop diagram—here, for the possibility of global warming or, more appropriately, climate change.

In analyzing causal loop diagrams, one seeks to determine if they are *reinforcing* or *balancing* loops. A reinforcing loop has an even number of negative links, and tends to lead to exponential increases or decreases, depending on whether the effects are positive or negative, respectively. A balancing loop has an uneven number of negative links, and tends to lead to a plateau in the affected variable, factor, or event. Identifying whether a causal loop is reinforcing or balancing is important to assess dynamic behaviors in a process or system (Figure 29–2).

The second type of causal diagram is a *fishbone diagram* or *herringbone diagram,* so named for its shape resembling the side view of the skeleton of a fish (Figures 29–3). Used since the 1940s, these cause-and-effect diagrams experienced a dramatic growth in the 1960s in Japan, where Kaoru Ishikawa pioneered their use in quality management processes, for which he became known as the proverbial father. Ishikawa proposed the diagram as a powerful quality management tool in Kawasaki's shipyard

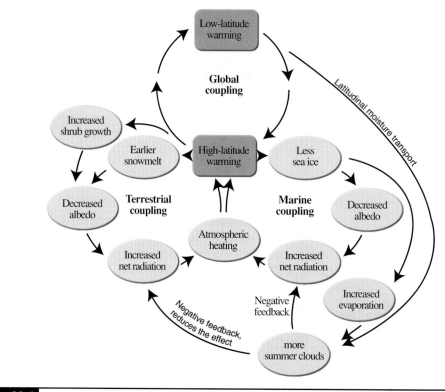

Figure 29-1 A causal diagram for the possibility of global warming, showing various couplings. (*Source:* http://maps.grida.no/go/graphic/climate-feedbacks-the-connectivity-of-the-positive-ice -snow-albedo-feedback-terrestrial-snow-and-vegetation-feedbacks-and-the-negative -cloud-radiation-feedback, drawn by cartographer Hugo Ahlenius for UNEP/GRID-Arendal, used with permission.)

and, later, in Mazda's automobile factories. The old fishbone diagram became known around the world as an *Ishikawa diagram.*

The most common uses of the *Ishikawa diagram* are for product or process design and for defect prevention in the pursuit of quality in production processing and manufacturing. The diagram helps identify potential factors that cause a particular effect (e.g., defects), each cause being a source for variability in output.

The categories of factors typically used in assessing manufacturing processes are called "the 8Ms," these being[1]:

1. Machine (i.e., technology)
2. Method (i.e., process)
3. Materials (i.e., raw materials, supplied parts, and information)
4. Manpower (including both physical and brain work)
5. Measurement (i.e., inspection)
6. Milieu/Mother Nature (i.e., environment)
7. Management/Money power
8. Maintenance

[1] Service industries tend to use "the 8Ps"—Product, Price, Place, Promotion, [Important] People, Processes, Physical Evidence, and Productivity and Quality—or "the 4Ss"—Surroundings, Suppliers, System, and Skills.

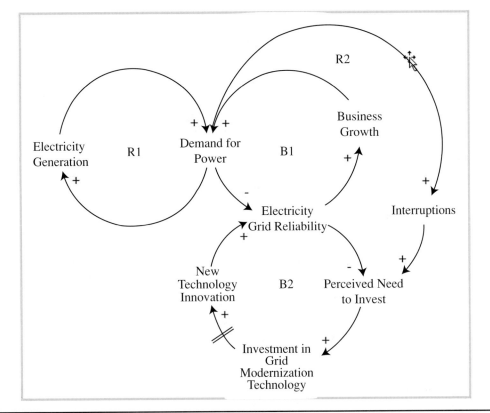

Figure 29-2 A generic causal loop diagram showing positive and negative effects. Each feedback loop has a polarity characteristic that can be determined by breaking the loop and tracing the path to see the direction of the effect. (*Source:* Jeff Wasbes at aec365.org; used with his permission.)

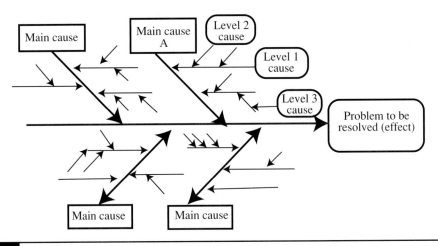

Figure 29-3 A generic fishbone diagram.

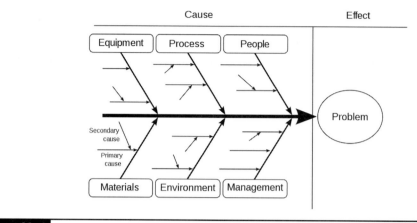

Figure 29-4 An Ishikawa diagram prepared to troubleshoot an errant process. (*Source:* Wikimedia Commons; originally contributed by slashme on 8 April 2008, and freely used under GNU Free Documentation License 1.2 or higher.)

Such a cause-and-effect diagram can, for example, reveal relationships among the 8M variables as well as provide insight into process behavior.

Illustrative Example 29-1: Using an Ishikawa Diagram

Just as flowcharts can assist in the analysis of a failure to pinpoint a root-cause (see Chapter 28, Illustrative Example 28–1), so, too, can Ishikawa diagrams. While the flowchart allows the investigator to determine at which step in design, fabrication, assembly, construction, or operation a failure is likely to have originated, one cannot be certain from a flowchart alone whether there is a possible or likely causal relationship between any particular step on the flowchart and the quality and performance of the final product.

By identifying the various factors that *could* influence a part, product, or process, an Ishikawa diagram can aid in further pinpointing the root-cause of the failure.

The best approach for creating an Ishikawa diagram for troubleshooting is to use a team. The procedure involves the following steps:

1. Have the team come to consensus on the clear and unambiguous statement of the problem to be addressed and write this statement in the box at the head of the Ishikawa diagram on a large whiteboard for all to see.
2. Jointly identify, then draw, and label the cause categories for major branches on the diagram.
3. Conduct a *brainstorming session* (see Chapter 31) in which each member of the team gives *one* (and only one) possible cause for a failure, and agree as a group where that cause should be placed on a branch. Label the cause. Repeat the process until every member of the team "passes," as ideas run out.
4. Interpret the Ishikawa diagram once completed to narrow down possible sources of the problem or failure.

Figure 29–4 shows an Ishikawa diagram useful for troubleshooting an errant process by considering primary and secondary causes relating to the process, equipment, materials, people, the environment, and management.

In Summary

Causal diagrams may be *loop diagrams* or *fishbone diagrams*—also known as *Ishikawa diagrams*—which visually/graphically depict the cause-and-effect relationships among factors in manufacturing, for example, and output, often measured in terms of quality and/or quantity. They can be invaluable techniques or tools in problem-solving.

CHAPTER 30

Decision Trees

THE THIRD AND FINAL major type of diagram used in engineering (along with flowcharts and causal diagrams) is the *decision diagram* or *decision tree,* so named for its treelike branches representing different decision options. The use and value of a *decision tree* are to identify (and, perhaps, model, in simulation) decision points and options and aid in the assessment of possible consequences. Consequences may include functionality, performance level, resource costs, chance-event outcomes, and others. In that they portray decisions, decision trees are another way (as opposed to flowcharts) for displaying algorithms.

Decision trees are widely used in design to identify and choose among alternative concepts during the conceptual stage. They also have value in problem-solving involving failures (e.g., in failure analysis investigations), in troubleshooting (e.g., in errant processes), and for aiding in a number of business decisions.

The convention for decision tree construction is to use squares to represent *decision nodes* (or decision points), circles to represent *chance-event nodes,* and triangles to represent *end nodes,* or final outcomes or output. The diagram is created (mentally) and drawn from left to right or top to bottom. It has only splitting paths (or *burst nodes*) and no converging paths (or *sink nodes*). Decision trees can grow to be quite large and are usually created manually (at least at first!).

Figure 30–1 shows a generic decision tree. In this generic example, the convention has been followed. The squares (1 and 2) represent decisions that need to be made. The lines that come out of a square at its right (or from the bottom for a top-down decision tree) show all available unique and distinct options that could be selected. The circles (A, B, and C) represent various circumstances or events that have an

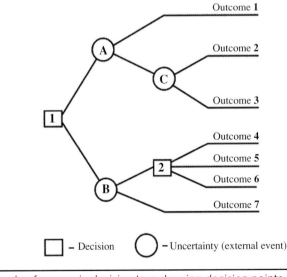

Figure 30-1 An example of a generic decision tree showing decision points (as squares) and circumstances for which the outcome is uncertain (as circles).

uncertain outcome (e.g., some problem that arises along that path). The lines that come out of circles at their right (or bottom) indicate possible outcomes for that uncontrollable circumstance. It is not at all unusual for best-guess probabilities for each such outcome to be written down to aid in decisions.

Each path, from left to right (or for top-down formats, top to bottom), leads to some specific outcome. Each of these end results needs to be described in terms of whatever criterion/criteria will be used to choose one outcome (and path) over another.

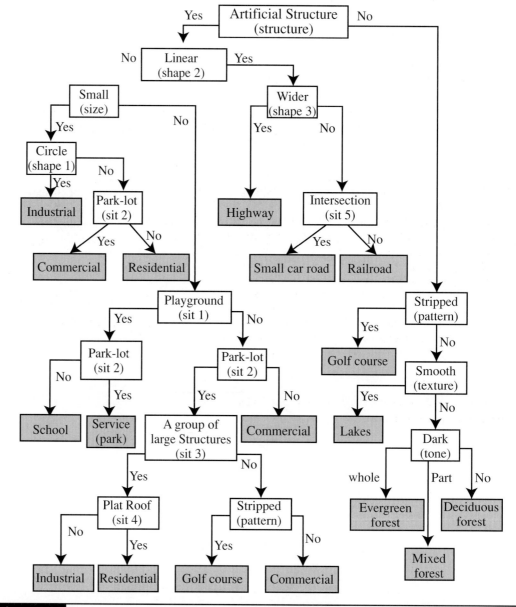

Figure 30-2 The top-down decision tree to help interpret urban use ground features using a rule-based system of questions and answers. (*Source:* "A Rule-based Expert System Using an Interactive Question-and-Answer Sequence," report by Jinmu Choi, formerly at the Department of Geography at the University of Georgia and now associate professor at Kyung Hee University in Seoul, Korea. Used with Dr. Choi's permission.)

It is good practice in engineering—and in life—to consider that each fork in a path involves *two* conscious decisions, not one, as it would seem. The first is why one road is chosen over the other. The second is why the other path was *not* chosen. For example, if you come to a fork in the road on your way home, you would choose one road over the other if only that choice leads to your home. However, if both roads would get you home, you might choose one over the other because that road is shorter and quicker. But, you might choose the longer, slower route if the shorter, quicker one is closed because a bridge is out! Considering both the positive reason for a choice and the negative reason for rejection can be useful later on if you have to reflect back on a failed path.

Besides their use in design in engineering, decision trees are used in teaching, albeit primarily in business schools.

Decision trees offer the advantages of (1) being simple to understand and interpret and (2) allowing insights to be gained even with little hard data. A disadvantage of decision trees is that they tend to bias users toward factors that include a large number of levels or options. This may or may not be warranted.

The key to a successful decision tree—and problem solution—is to initially include all options at each decision point, leaving critique of one option over another to later in the process. The most creative—and innovative—designs may evolve from nontraditional options.

Illustrative Example 30–1: Using a Decision Tree in Problem-Solving

Interpretation of aerial photographs is frequently aided by the use of computers and techniques that use pattern recognition. Some techniques employ top-down decision trees and a series of questions and answers in a rule-based system of artificial intelligence (AI). Seven elements are used to interpret features shown in the photograph being interpreted: time, texture, pattern, size, shape, situation, and shadow. Only the first six are used in the AI-based decision tree shown in Figure 30–2, as shadow is generally too small to detect in aerial photographs.

Using the six elements listed here, rules in a rule-based system allowed urban uses to be distinguished from natural features such as lakes and forests.

In Summary

Decision trees provide a convenient, easy-to-understand visual display of the points in a design where a decision among options is required. They are a powerful problem-solving technique in design.

PART FOUR

Conceptual or Abstract Approaches to Problem-Solving

Brainstorming: Process, Methodology, and Technique

BY NOW, after reading Parts 1, 2, and 3, it should be clear that there are problems whose solution approach and solution either clearly demand or would benefit from the use of a mathematical approach (Part 1), a physical/mechanical approach (Part 2), or a graphical approach (Part 3). All of these approaches have in common the use of a physical or sensory trigger to direct abstract thinking and mental concepts.

A mathematical equation that represents a physical system (e.g., $F = ma$, $Q = k \, dT/dx$, $D = D_o$ exp $(-Q_d/RT)$ is not abstract. It simply requires values (numbers) to be put in for each known term in order to calculate an unknown term by plug-and-chug (Chapter 3). A physical model of a crank-and-arm mechanism or an arch simply requires application of a force or action to see the reaction (e.g., see Chapter 18). A graph of measured data suggests the form of the governing equation (e.g., data points that lie on a line with a slope that identifies the power-law relationship between input(s) and output (Chapter 26).

Here, in Part 4, the focus is on approaches and/or techniques for solving problems that begin with mental abstractions of the problem and mental concepts for solving the problem. In other words, for the approaches and/or techniques addressed hereafter, an abstract idea or concept triggers an action or actions to solve the problem, as opposed to sensory inputs and physical manipulations of an equation, a physical model, or a graphic image triggering an abstract thinking process as the action.

Some of the conceptual or abstract (mental) thinking approaches to problem-solving that are discussed hereafter require the use of mathematics in pursuing possible solutions (e.g., problem dissection, with decoupling and coupling, in Chapter 33, and backward problem-solving and inverse problems in Chapter 34). Most, however, rely on visual, graphical, or iconic methodologies. None explicitly requires the use of a physical/mechanical model.

So let's begin with the conceptual approach of *brainstorming*.

Did you ever hear someone say—or say yourself—"I just had a brainstorm"? Or maybe someone—or you—said, "It came to me like a bolt out of the blue," as reference to a lightning bolt from the sky? If you did, you will understand the mental act and appreciate the mental exercise of *brainstorming* as a way of coming up with a solution or bunch of possible solutions to a problem (Figure 31–1).

Brainstorming is a technique for eliciting creative thinking by which a group, or even an individual, tries to find a solution for a specific problem by assembling a list of ideas spontaneously contributed by members, or that cascade from an individual's mind.[1] First popularized by Alex Faickney Osborn in his book *Applied Imagination* in 1953, the method originated in 1939 when Osborn began hosting group-thinking sessions to develop creative ideas for ad campaigns when none were forthcoming from individual workers. He claimed that a significant improvement in the quantity and quality of ideas

---

[1] Everyone has experienced the exhilarating instances when ideas pop into one's head faster than they can be expressed or written down, with one idea triggering another, and so on and so forth. This experience is referred to as "a stream of consciousness."

Figure 31-1   A simple, but vivid, graphic showing the central aspect of the problem-solving technique of brainstorming, that is, that ideas enter participants' minds like lightning bolts from the clouds.

by employees occurred with the group approach *provided* four principles were adhered to, with the intention of reducing social inhibitions among group members. These four principles are:

1. *Focus on quantity.* The assumption is that the greater the number of ideas generated, the greater the chance of producing an effective and/or radical solution.
2. *Withhold criticism.* By suspending judgment, participants will feel free to express unusual ideas.
3. *Welcome unusual ideas.* These can be generated by looking at a problem from different and new perspectives and by temporarily suspending assumptions—and criticism.
4. *Combine and improve upon ideas.* It is believed that ideas are frequently generated by a process of association and that good ideas may be combined to form a better idea.

To work best, brainstorming should address a simple specific question.

Research on brainstorming brings into question whether groups are consistently capable of outperforming a group (or set) of individuals' collected ideas. "Real" groups often produce fewer creative ideas than "nominal" groups in which ideas are generated independently of one another by different people working on the same problem remotely from, but sharing with, one another. Some inhibitors are suggested to be at play in real groups, including: (1) *free riding* (i.e., not liking for one's idea to be combined with other ideas), (2) *evaluation apprehension* (i.e., fear of eventual criticism— even though put on hold during idea-generating sessions), and (3) *blocking* (i.e., only one person being able to gainfully voice his or her ideas in a group at a time). The adoption of *group support systems,* wherein individuals submit ideas on a shared computer space to allow instant and anonymous visibility to an entire team, seems to remove blocking effects and seems to allow significantly greater creativity to emerge.

Group Techniques

There are a variety of techniques for group brainstorming. These include the following in bulleted summaries:

Nominal Group Technique:
   1. All participants have an equal say.
   2. Participants write down ideas on a piece of paper independently.
   3. A facilitator collects and posts ideas as they are generated.
   4. Ideas are distilled by a hand vote to allow rank ordering.
   5. Top-ranked ideas may be sent back to the group or subgroups for further brainstorming.

Group Passing Technique (alternatively, Brainwriting):
   1. Each person in a group arranged in a circle writes down one idea on a piece of paper.
   2. The pieces of paper—and ideas—are passed to the next person, in a clockwise direction, whereat some new thoughts might be added.
   3. This continues until each originator gets back his or her original piece of paper.
   4. A facilitator collects all pieces of paper and records the ideas and thoughts for all to see and for the leader to consider.

Team Idea Mapping:
   1. Each participant brainstorms a well-defined question or problem individually.
   2. All ideas are then merged onto one large map to group similar ideas, show linkages, and so forth, and to stimulate new ideas.
   3. Once all ideas are captured, the group can prioritize and take action.

Electronic Brainstorming:
   1. With an electronic meeting system, participants share a list of ideas over the Internet, with ideas having been generated independently by individuals in the "nominal" group.
   2. If users are able to log on over an extended period of time (typically, one or two weeks), there is time for ideas to "soak" and opportunity for feedback.

Figure 31–2 schematically summarizes the general process of brainstorming.

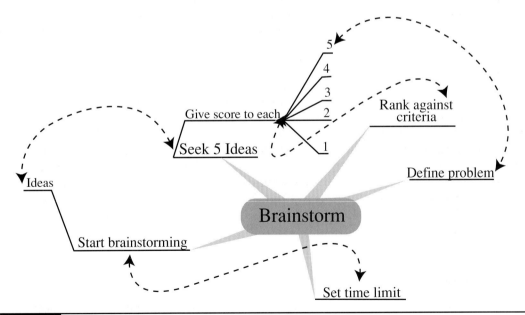

Figure 31-2   A schematic illustration of the process flows during a group brainstorming session.

Individual Brainstorming

Brainstorming can be done by an individual operating in solitude. The process is particularly useful for writing papers but can also be adapted to problem-solving. To elicit ideas, there are several techniques that can be considered, as follows:

Use analogies. Think of how the problem being faced resembles a related—even remotely related—experience from your past or a problem solved by you or someone else in the past. This technique is dealt with in detail in Chapter 32.

Use freewriting. Put down your thoughts—any and all thoughts—as fast as you can. Let the thoughts flow. Don't worry about the quality of the idea, your writing style, or surface-level issues such as spelling, grammar, or punctuation. Talk—write—to yourself. If you can't think of what to say, write that down! Set a time limit—say, 15 minutes. When you're done—or the allotted time has expired—read back over what you have written. Find the "gems" and highlight them, organize them, and build upon them.

Think outside the box by using fantasy. Remove all constraints on your thinking. Try to imagine what you would like to have if there were no limits imposed on you. Try to think about your problem from different perspectives, perhaps by imagining you are someone else.

Use shapes, sketches, charts, graphs, or tables. Forget words! Let images drive your thinking. Capture your ideas as simple graphic images (e.g., shapes or sketches), as graphs or plots, or as tables. Later, once ideas have been generated and captured, write a note to yourself about each; in other words, annotate your nonverbal ideas.

Individual brainstorming is for you, no one else. Be free and say or write or graphically express what comes into your mind. It turns out that despite the rage in academe—due to the rage within funding agencies—for collaborative research, the truly innovative ideas still come from "lone wolves." Without social pressure, creativity flows best! But don't wait to be inspired. You are likely to wait a long time—to no avail. Good work comes from hard work—as my father said.

Illustrative Example 31–1: A Step-by-Step Guide to Group Brainstorming

A proper group brainstorming session requires a facilitator to capture ideas—without any vested interest or bias—and a space to work and a place to write or post ideas, such as a whiteboard or flip chart. Post-it Notes work well.

The process works best with a varied group of 8 to 12 people working in a relaxed environment (free of telephones, cell phones, pagers, e-mail, etc.). Creative exercises, relaxation techniques, or fun activities can help participants relax their minds and become more comfortable with one another.

The steps are then:

1. Define the problem or issue as a concise creative challenge. For example: "In what ways might we . . .?" or "How could we . . .?"
2. Set a time limit—preferably about 20 or 25 minutes (so as not to let people get bored). Larger groups tend to need more time to allow everyone to express their ideas. Alternatively, set an "idea limit," aiming for a minimum but accepting and welcoming more. How many depends on the sizes and complexity of the problem or issue.
3. Once the brainstorming begins, participants should shout out ideas or solutions while the facilitator writes them down for all to see. There can be *no* criticizing of ideas! It doesn't matter how silly, impossible, or weird the idea is. Laughing is okay. Criticism is not!

4. When the time limit is up, select the top five ideas by consensus. (The originator of an idea can expound upon his or her idea at this point to clarify or defend it.) If six or seven ideas are close, keep them all. Narrow the list later.
5. As a group, by discussion and consensus, write down four or five criteria for judging which ideas best solve the problem or address the issue at hand. Start each criterion with "It should be . . . ," for example, ". . . legal," " . . . cost effective," " . . . easy to accomplish," and so on.
6. Give each idea a score of 0 to 5 points (maximum) depending on how each meets each criterion. Once all the top ideas have been scored against each criterion, add the scores for all criteria for each idea. (It is sometimes worthwhile to assign a greater weighting factor to one criterion over another; for example, "It should provide the greatest functionality" would have greater weight than "It should be easy to accomplish" or even "It should be cost effective." After all, what good is cost effectiveness if the functionality is not there?
7. The idea with the highest score will be the best solution to the problem. *But* keep a record of all the best ideas and their scores in case the best idea proves to be unworkable.[2]

A particularly good treatment of brainstorming, along with other *concept generation* techniques is given in Chapter 9 ("Concept Generation") of Barry Hyman's *Fundamentals of Engineering Design*.

In Summary

Brainstorming is a popular and potentially powerful technique for eliciting creative ideas, whether performed by a real or a nominal group working face-to-face or in a virtual space, respectively, or by an individual working in solitude. Even though there has been—and remains—some debate about the true productivity of one method compared to another, the process is widely used and is always worth a try.

Suggested Resources

Hyman, Barry, *Fundamentals of Engineering Design,* Prentice-Hall, Upper Saddle River, NJ, 1998.

---

[2] A good strategy adopted by the author when he worked as an engineer was to take a two-pronged approach to problem-solving, with one approach focusing on a surefire solution and the other on a higher-risk approach. The surefire approach would absolutely yield a solution to the problem, albeit perhaps not an earthshaking one with tremendous payoff potential. The higher-risk approach would have the potential for tremendous payoff if it could be pulled off. The idea was: Pleasantly surprise management when one can, but never let them down!

CHAPTER 32

Using Analogs

WHEN YOU LEARNED about the structure of an atom, you almost certainly were shown the planetary model proposed by Niels Bohr, in which tiny negatively charged electrons circulate around a net-positive charged nucleus in an *analogy* to the way the planets in our solar system circulate around the Sun in orbits (Figure 32–1). In fact, the analogy is sometimes extended to relate the scale for the radii of the various electron orbits to the radii of the nucleus to the scale for the radii of the orbits of the planets to the radius of the Sun, making the point that most of the atom, like most of the solar system, is empty space.[1]

The word *analogy* (from the Greek *analogia,* "proportion") is defined as something like (with the author's own twist): a structure, entity, or system in one context that is similar in appearance or behavior to one in another context from which the former may or may not be derived.

Analogy plays a significant role in problem-solving, as well as in perception, memory, explanation, creativity, and communication, as it allows associations to be made and linkages to be identified. It has been argued that analogy is "the core of cognition." In the cognitive process an analogy transfers meaning or information from one subject (or source) to another subject (or target). Analogies can be used for giving examples, making comparisons, or, in storytelling, to convey allegories or parables. Analogies are powerful aids in finding and seeing associations between what may not be obvious concepts, events, factors, behaviors, processes, and so on.

Analogy has been used since ancient times, most notably in ancient Greece, where it was used in the mathematical sense to mean "proportionality." It evolved to be understood as an "identity of relation" between any two ordered pairs, whether mathematical in nature or not; for example, *palm* is to *hand* as *sole* is to *foot.*

Analogy took a great leap forward when Greek philosophers, such as Plato and Aristotle, saw it as

---

[1]It has been calculated that if all the electrons, protons, and neutrons for all the atoms making up everything on Earth, including the earth, were combined, the volume would occupy only a few cubic centimeters.

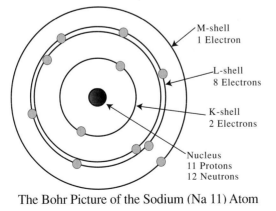

The Bohr Picture of the Sodium (Na 11) Atom

M-shell
1 Electron

L-shell
8 Electrons

K-shell
2 Electrons

Nucleus
11 Protons
12 Neutrons

Figure 32-1 A schematic of Bohr's planetary model of the atom; here, for sodium (Na), atomic number 11.

a "shared abstraction." By this concept, analogous objects share not only a relation but also an idea, a pattern, a regularity, an attribute, an effect, or a function. An analog watch, for example, relates the position of long and short hands on the dial to represent the time on a 12-hour cycle or arc, even if there are no numerals to help. In fact, the analogy of the watch dial was extended to refer to position in either a horizontal or a vertical plane (e.g., "Enemy aircraft at 11 o'clock!").

The power of analogies or analogs in problem-solving in engineering is that they give engineers insight. A difficult concept is frequently made clearer and more understandable by saying, "The analogy is . . . ."

Analogies are specifically used in three major generic ways in engineering:

- Thermal-electrical analogs
- Electrical-mechanical analogs
- Biological analogs

There are surely more than three, but these three are most common and serve to make the desired points.

Illustrative Example 32–1: Using a Thermal-Electrical Analog

A former colleague, Dr. Kevin Craig (now at Marquette University as the Robert C. Greenheck Chair in Engineering Design and Professor of Mechanical Engineering in the College of Engineering), in his paper "Insight Based on Fundamentals Is the Key to Innovation in Multidisciplinary Problem Solving" (Electronics, Design, Strategy, News, at www.edn.com, March 01, 2010), gave this splendid example (used with his kind permission):

> Consider an exhaust system of a motorcycle and its heat shield. Temperatures have to be controlled through design for performance but also to protect the rider. Being able to model this system as a network of thermal resistances and capacitances, just like an electrical circuit, is a powerful design tool. It allows the engineer to visualize the flow of heat and the storage of thermal energy, and specify key temperatures by selection of materials and geometries that vary the network thermal resistance (conduction, convection, radiation) and capacitances. Improving performance happens with understanding—not by trial and error—and quickly.

Electrical-Mechanical Analogs

An extremely valuable analog of great utility in engineering problem-solving is that between electrical and mechanical systems (Figure 32–2). In fact, there are two analogs that are used to go between electrical and mechanical systems. One (I) uses the analogy between the force and the current, while the other (II) uses the analogy between the force and the voltage. Table 32–1 shows analogous quantities between the systems for the two types.

Measurements of current or voltage (depending on the specific analog type, I or II, respectively) allow easy conversion to force in the mechanical system represented by that circuit.

The real key to the electrical-mechanical analog is that analogous electrical and mechanical systems are governed by differential equations of the same form (see Chapter 3, "Equation Forms"). Table 32–2 gives the analogous equations.

The important relationship when converting from an electrical circuit to a force-current mechanical analog (Type I), for example, is that between Kirchoff's current law and D'Alembert's law[2]:

$$\Sigma_{\text{node}} \text{ current} = 0 \Leftrightarrow \Sigma_{\text{object}} \text{ forces} = 0$$

[2]This analogy works easily only for capacitors that are grounded.

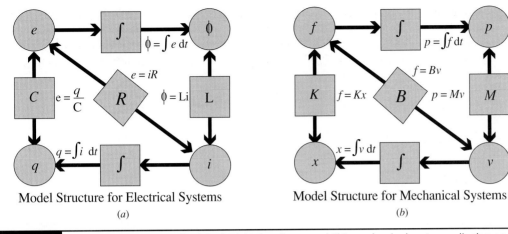

Model Structure for Electrical Systems
(a)

Model Structure for Mechanical Systems
(b)

Figure 32-2 The model structures for (a) electrical systems and (b) mechanical systems display an obvious analogy. (*Source:* Kevin C. Craig, "Insight Based on Fundamentals Is the Key to Innovation in Multidisciplinary Problem Solving," Electronics, Design, Strategy, News, at www.edn.com, March 01, 2010; used with Professor Craig's permission.)

The important relationship when converting from an electrical circuit to a force-voltage mechanical analog (Type II), on the other hand, is that between Kirchoff's voltage law and D'Alembert's law, with inertia forces included:

$$\Sigma_{\text{loop}} \text{ voltage} = 0 \Leftrightarrow \Sigma_{\text{object}} \text{ forces} = 0$$

The beauty of this analogy is that conversion can be done in either direction, that is, electrical to mechanical or mechanical to electrical.

Details on this powerful problem-solving tool can be found on the excellent website at http://lpsa. swarth/ElectricalMechanicalAnalogs.html by Erik Cheever.

As an interesting historical note, the earliest electronic computers operated using analogs, known as *analog computers*. Circuitry in these computers (which began with rectifying and amplifying vacuum tubes analogous to p-n junction rectifiers and p-n-p or n-p-n transistors) allowed the mathematical operations of addition, subtraction, multiplication, division, differentiation, and integration to be performed using circuitry. Answers for the mathematical operations were obtained from measured voltages.

TABLE 32-1 Analogous Quantities between Electrical and Mechanical Systems

Electrical Quantity	Mechanical Analog I (Force-Current)	Mechanical Analog II (Force-Voltage)
Voltage e	Velocity v	Force f
Current i	Force f	Velocity v
Resistance R	Lubricity $1/\mu$ (inverse friction)	Friction μ
Capacitance C	Mass M	Compliance $1/K$ (inverse spring constant)
Inductance L	Compliance, $1/K$ (inverse spring constant)	Mass M
Transformer $N_1{:}N_2$	Level $L_1{:}L_2$	Lever $L_1{:}L_2$

Source: After http://lpsa.swarth/ElectricalMechanicalAnalogs.html by Erik Cheever; used with permission.

TABLE 32-2   Analogous Constitutive Relationships for Electrical and Mechanical Analogs

Electrical Equation	Mechanical Analog I (Force-Current)	Mechanical Analog II (Force-Voltage)
$e = i/R$	$v = f/B$	$f = vB$
$e = L\,di/dt$	$v = 1/K\,df/dt$	$f = M\,dv/dt = M.a$
$i = 1/L\int e.dt$	$f = K\int v.dt = K.t$	$v = 1/M\int f.dt$
$e = 1/C\int i.dt$	$v = 1/M\int f.dt$	$f = K\int v.dt = K.x$
$i = C\,de/dt$	$f = M\,dv/dt = M.a$	$v = 1/K\,df/dt$
power $= e.i$	power $= v.f$	power $= f.v$
Transformer $e_1/e_2 = N_1/N_2 = i_2/i_1$	Lever $v_1/v_2 = L_1/L_2 = f_2/f_1$	Lever $f_1/f_2 = L_2/L_1 = v_2/v_1$
capacitor energy $(1/2)C.e^2$	mass energy $(1/2)M\,v^2$	spring energy $(1/2)K.x^2 = (1/2K(f/K))^2 = (1/2)\,f^2/K$
inductor energy $(1/2)L.i^2$	spring energy $(1/2)K.x^2 = (1/2K(1fK))^2 = (1/2)\,f^2/K$	mass energy $(1/2)M.v^2$
Σ_{node} current $= 0$	Σ_{object} forces $= 0$	Σ_{loop} velocities $= 0$ (not explicitly used)
Σ_{loop} voltage $= 0$	Σ_{loop} velocities $= 0$ (not explicitly used)	Σ_{object} forces $= 0$
voltage of ground $= 0$	velocity of ground $= 0$	???

Source: After http://lpsa.swarth/ElectricalMechanicalAnalogs.html by Erik Cheever; used with permission.

Biological Analogs

Biological entities and systems have intrigued and inspired human beings since their appearance on Earth. There can be little doubt that human beings were inspired to fly by seeing birds fly and watching shorebirds joyfully soar on thermal updrafts. Greek mythology tells us that Daedalus and his son, Icarus, inspired by birds, stuck feathers to their arms using beeswax. Daedalus warned his son not to fly too close to the Sun so as not to have the beeswax melt. But the overzealous boy did fly too close to the Sun, whereupon the beeswax melted and he plummeted into the sea (Figure 32–3).

Whether based on truth or not, Icarus's plight did not deter other real human beings from attempting flight using birdlike wings, which they flapped after jumping from a cliff or running along the ground.

Many elegant solutions to engineering problems have been inspired by biological phenomena. In all likelihood, many more will be similarly inspired in the future. In fact, *biomimetic design,* which has attracted newfound interest, involves design that copies biological models. The logic for doing this seems overwhelming. Mother Nature's designs are, themselves, elegant in their simultaneous simplicity and sophistication. More so, such designs are almost always the result of evolution that sorts out successful designs from unsuccessful ones. Many biological designs are also self-correcting—for example, from Purves et al. (see "References" under "Suggested Resources"):

Blood must be returned from veins to the heart so that circulation can continue. If the veins are above the level of the heart, gravity helps blood flow, but below the level of the heart, blood must be moved against the pull of gravity. If too much blood remains in the veins, then too little blood returns to the heart, and thus too little blood is pumped to the brain; a person may faint as a result. Fainting

Figure 32-3 Painting showing Daedalus and his son, Icarus, from Greek mythology, flying using bird feathers attached to their arms using beeswax. Icarus is seen falling into the sea after flying too close to the sun so that the beeswax melted. (*Source:* http://revolutioncontinues .com/wp-content/uploads/Daedalus-and-Icarus.jpg; used with permission.)

[however] is self-correcting: A fainting person falls, thereby moving out of the position in which gravity caused blood to accumulate in the lower body.

Modern engineers should be vigilant and observe and learn from nature. They may be inspired for some design, but, if not, at least they got to enjoy life more by watching butterflies fly and flowers turn their faces toward the sun.

Illustrative Example 32–2: Using a Biological Analog

All birds are able to reshape (i.e., morph) their wings to optimize flight characteristics for different situations. This is especially noticeable—and pronounced—in raptors, such as America's proud symbol, the American bald eagle, as well as red-tailed hawks such as shown in Figure 32-4. Such large birds cup their wings forward and curve (or "gull") them downward to increase drag (by increasing the angle of attack) and slow forward velocity to induce stall to allow pinpoint landing. To soar as it hunts for prey on the ground, it extends its wings to the fullest span, tip feathers splayed for even greater lift and to provide subtle movement for maneuverability. To attach its prey on the ground, it sweeps and tucks its wings to reduce drag and allow high-speed dash.[3]

[3]When a fighter aircraft with variable-geometry wings, like the Grumman F14, swept its wings, its center of lift shifted aft from its center of mass or gravity, which required compensation by the pilot to maintain level flight. In the same way, large birds of prey, such as eagles, hawks, and falcons, hunch their inner wing forward as they sweep their outer wing aft. This elegant motion keeps the center of lift and center of mass coincident, increasing stability.

Figure 32-4 Photographic montage showing how a red-tailed hawk changes the configuration of (i.e., morphs) its wings to optimize aerodynamic performance in different flight regimes. (*Source:* Image courtesy of photographer Ron Reznick at www.Digital.Images.net.)

Enabled by advances in flexible materials, smart materials, and computer-based flight control (to deal with instabilities during transitions while morphing), NASA is exploring morphing wing technology for future aircraft, particularly unmanned, pilotless aircraft (Figure 32-5). The similarity between what man-made morphing-wing aircraft and Mother Nature's magnificent fliers do is clear.

The subtle lesson: As materials and other technologies have advanced, flying machines are being made more birdlike in a closer mimic of Nature—which got it right.

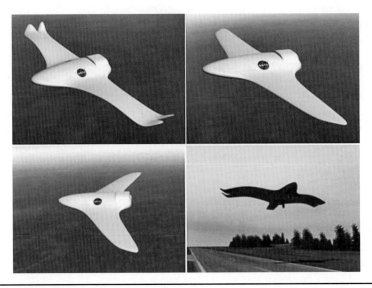

Figure 32-5 Images of NASA's exploration of morphing-wing aircraft. (*Source:* NASA files; no permission required.)

In Summary

Analogs help one make associations that may not be obvious. Analogs are at the core of our cognition, facilitating perception, memory, explanation, creativity, and communication. In engineering, common analogs exist and are used between thermal and electrical systems, electrical and mechanical systems, and arising from biological systems.

Suggested Resources

References

Purvis, W. K., D. Sadava, G. H. Orians, and H. C. Heller, *Life: The Science of Biology,* 6th edition, Sinauer Associates, Sunderland, MA, 2001.

Related Readings

Steadman, Philip, *The Evolution of Designs: Biological Analogy in Architecture and the Applied Arts,* revised edition, Routledge, New York, 2008.

Wormeli, Rick, *Metaphors and Analogies: Power Tools for Teaching Any Subject,* Stenhouse Publishers, Portland, ME, 2009.

CHAPTER 33

Dissecting Problems: Decoupling and Coupling, Zooming In and Zooming Out

SOME PROBLEMS ARE SO BIG and/or so complex, they are intimidating or overwhelming or seemingly impossible to solve. Imagine being part of a small group of three or four people presented with an elephant and being told the task is to eat the entire elephant! Many, if not most, such small groups would simply give up, saying it is impossible to eat an entire elephant. But a small group of good engineers—or very good engineering students—would, after the initial shock wore off, set about dissecting the elephant to attack the task bit by bit—or bite by bite, as it were. They might start from the trunk or start from the tail, but they would eat that elephant!

Real-world engineering is filled with big and/or complex problems that need to be solved. (In fact, these are often career makers or breakers.) Building a bridge across a deep rock ravine. Sending a man to the Moon and getting him back to Earth safely. Finding ways to provide the people of the world fresh drinking water, clean, renewable energy, and the quality of life they all seek and deserve. These are big problems, with many facets and involving complex interacting factors and phenomena and events.

The way to tackle big and/or complex problems is to dissect them into logical and more manageable parts and then set about systematically attacking each part—part by part, facet by facet, bit by bit. This *dissection of problems* involves two phases: (1) to dissect the overall problem into both logical and manageable (hopefully, solvable) parts and (2) to combine the solutions for each part into an integrated and integral whole. For problems involving multiple phenomena occurring at the same time and in the same places, dissection usually involves *decoupling* the individual phenomena to allow each to be addressed independently, based on the assumption (see Chapter 3) that they *are* independent of one another or not, and, then, *coupling* the independent solutions to re-create or reconstruct the entire situation, system, or event.

The processes of *decoupling* and *coupling* are nontrivial for several reasons. More often than not, the various phenomena involved interact with one another to some degree in ways that make simple addition of each insufficient (see Chapter 11). In other words, the whole may be something different than the simple sum of the parts. In some cases, one phenomenon interacts with another phenomenon such that there is some nullification or even antagonistic interaction. In other cases, one phenomenon interacts with another phenomenon with a reinforcing effect that is synergistic, the two together resulting in something greater than expected from simple addition. (This is very common in metallurgy, for example, as when both chromium and molybdenum are added to a carbon steel, with the combination of the two solutes having a greater effect on the ease with which the steel can be hardened by rapid cooling than the sum of the effect of each alone.)

Rather than continue to talk in generalities, let's look at some specific (albeit not all-inclusive) examples for which decoupling is necessary and coupling may be more complicated than simply adding the effects of each part. Four illustrative examples should suffice to make the point, one involving interacting forces, one involving interacting stresses or fields, one involving interacting energies, and one involving interacting thermal, mechanical, and metallurgical effects.

Illustrative Example 33–1: Interacting Forces

There are many instances and situations in science and in engineering in which forces that arise from different sources interact. A good example is the convective flow (i.e., flow due to convection) that occurs in the molten pool of fusion welds during their creation. Four forces contribute to convection: (1) a buoyancy force, (2) a surface-tension-gradient force, (3) an electromotive force, and (4) and an impingement force.

A *buoyancy force* arises from the effect of gravity on portions of the molten metal in the weld pool having different densities as a result of their being at different temperatures. The hotter liquid near the center of the weld pool (where the energy from the welding heat source is being deposited) is less dense, so it rises. The cooler liquid near the edges of the weld pool (where the pool meets the unmelted base metal) is more dense, so it sinks due to the force of gravity. This creates a slow (~1 cm/s) outward circulation.

Cooler molten metal (like all liquids) has a greater surface tension than hotter molten metal and, so, pulls molten metal from the hot center to the cooler edges of the pool, where the moving liquid is forced to turn downward. This *Maragoni force* creates a much faster (~10 to 100 cm/s) circulation in the same direction as the buoyancy force.

An electromotive force (emf) arises in electric welding processes (e.g., arc and electron-beam welding) due to the interaction between the magnetic field surrounding the current-carrying electrode (or beam) and the electrically conductive—and moving—molten metal in the weld pool. This *Lorentz force* creates a fairly fast (~10 cm/s) circulation in the opposite direction of the Marangoni and buoyancy forces.

The overall pattern of convective flow in a molten weld pool is the algebraic sum of these various forces, as vectors. Which force components act, and to what degree each acts, depends on the specific welding process (e.g., there is no Lorentz force component for oxy-fuel or for laser-beam welding). The interactions for both positive and negative surface tension gradient forces with buoyancy and electromotive forces in gas-tungsten arc welding are shown in Figure 33–1.

Illustrative Example 33–2: Interacting Stresses

Stresses that develop in mechanical systems or structures can arise from any or all of three generic sources. These three generic stress types are: (1) external stresses, (2) internal stresses, and (3) residual stresses (which are also an internal stress type). Sources of external stresses include applied or service loads from some driver (human beings, motors or engines, other structures or structural elements, etc.); wind, air (as drag), or water (or other fluid); friction (at moving interfaces); and thermal sources that cause dimensional changes.[1] Internal stresses arise from the force of gravity, inertia, or springs or things that act like springs by storing energy that can create a force. Another category of internal stresses is residual stresses. Residual stresses differ from other internal stresses in that they are locked in the structure and exist in the absence of any external loading or thermal gradients. Things that give rise to residual stresses are normally associated with processing, including gradients in elastic-plastic strain during deformation processes (e.g., forming and

[1] Temperature changes or differences (due to gradients) cause expansion on heating and contraction on cooling. Thermal stresses are external if dimensions are free to change but give rise to internal locked-in residual stresses if they are not free to change.

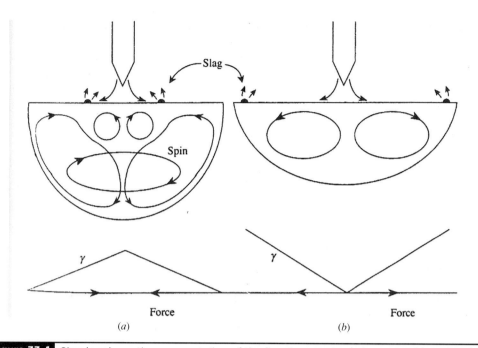

Figure 33-1 Simple schematic representation of the flow in a gas-tungsten arc weld pool under the combined effects of buoyancy and electromotive force for a surface tension gradient force that is (a) positive and (b) negative. (*Source:* J. F. Lancaster, *Metallurgy of Welding*, 5th edition, Chapman & Hall, London, 1993; used with permission of Kluwer Academic Publishers, The Netherlands.)

machining), shrinkage (e.g., from casting, molding, or fusion welding), or gradients in local specific volume due to heat-treatment microstructural changes. Residual stresses develop only if the material—locally—is not free to move in response to the stress, whatever its source.

Whenever a structural-mechanical system is being assessed, all external, internal, and residual stresses must be considered. The net stress pattern in a structure is the algebraic sum of these stresses (e.g., tension stresses being positive [+] and compressive stresses being negative [−]). It is not unusual in a failure analysis investigation to find cracks oriented differently than would be expected from known applied and internal stresses alone.[2] The culprit is usually the presence of a residual stress.

Illustrative Example 33-3: Interacting Energies

A good engineering student—and, hopefully, all engineers—know that "energy is energy." One form can be converted to another, and one form adds to another.

A simple, but important, example of this arises from the net potential energy curve for a crystalline solid (e.g., metal; Figure 33-2).

The depth of the potential energy curve at absolute zero (i.e., 0 K) represents the binding energy of atoms in the metal (or ions in ionic ceramics or molecules in polymers). It

---

[2]When a solid material fractures, cracks are oriented perpendicular to the major or net stress if the material acted in a brittle fashion and at ±45° to the major or net stress if the material acted in a ductile fashion. Thus, crack orientation is a key clue to the nature of the failure, that is, ductile or brittle fracture.

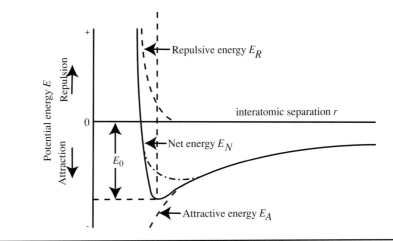

Figure 33-2 Schematic plot of the net potential energy curve between two atoms, ions, or molecules in a solid. The minimum net potential energy is where the atoms, ions, or molecules reach their equilibrium separation distance. The depth of the well corresponds to the energy needed to break the bond. The solid curve is for 0 K; the long-short-long dashed curve, for 300 K (*RT*).

would take energy equivalent to the binding energy to break the bonds between atoms (or ions or molecules). If this energy is entirely mechanical, the binding energy indicates the cohesive strength of the solid. If the energy is entirely thermal, the binding energy indicates the melting point of the solid. If the temperature of the solid is greater than absolute zero—say, 300 K (room temperature)—less mechanical energy is needed to break the bonds and pull the solid apart because some portion of the energy needed to do so is already present as thermal energy stored in lattice vibrations. This is why pure metals, for example, become progressively weaker as their temperature goes up.

In the case of corrosion, which involves the breaking of bonds to destroy the solid material by chemical energy provided by the corrosive agent, the rate at which corrosion takes place increases as the temperature goes up (from absolute zero), since some of the energy needed to break bonds is already present thermally. If, on top of this temperature effect, one adds mechanical stress (tensile or compressive, applied or residual, it doesn't matter!), the rate at which corrosion takes place is increased. The reason: The total energy to break bonds—being the binding energy—is the sum of thermal, chemical, and mechanical energy acting on the solid.

Illustrative Example 33–4: Thermal-Mechanical-Microstructure Interaction

A solder joint in a surface-mounted device[3] (Figure 33–3) must carry all of the stress imposed by all sources, mechanical and thermal, as there is no compliance from fine wire leads, which act like springs. Even presuming no mechanical stress is directly applied to the device—and, thus, to the solder joint—a thermally induced mechanical stress is present due to the differences in coefficient of thermal expansion (CTE) for the substrate (e.g.,

---

[3] Modern microelectronic devices generally involve surface-mount technology (SMT), in which a leadless device is soldered directly to deposited metal (e.g., Cu) on a circuit board, for example, as opposed to the older "through-hole" technology, in which fine wire leads were passed through holes in the circuit board and mechanically cinched as well as soldered. In such leaded devices, small loops were usually formed in each fine wire lead to act as a spring, preventing loads on the device from passing directly and entirely to the solder proper.

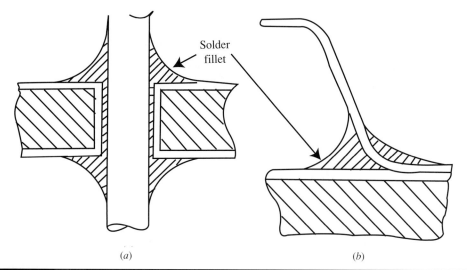

(a)   (b)

A schematic illustration of a surface-mounted versus a through-hole solder joint. In the leadless SMT joint, stress develops in the solder due to differential CTEs between the circuit board, chip package device, and solder alloy.

fiberglass circuit board), the device (e.g., ceramic chip package), and the solder (e.g., 60–40 Sn-Pb alloy). If the temperature environment in which the SMT device operates fluctuates up and down—due to I^2R heating, if not due to changing surroundings—differential CTEs give rise to fluctuating thermally induced mechanical stresses. The result is thermomechanical fatigue (TMF).

What occurs in the solder joint is complicated. The magnitude of the induced mechanical stress depends on the magnitude of the temperature change and the magnitude of the difference in CTEs. However, the strength (e.g., the yield strength and, therefore, the fatigue strength) of the solder depends on the temperature, decreasing as the temperature increases (per Illustrative Example 33–3). When the temperature is low, the solder may be below the point (approximately 0.4 absolute MP) where it would exhibit creep, that is, time-dependent strain under sustained stress. Hence, microscopic damage occurs here. When the temperature is high, the solder may be above the temperature where it will undergo creep and, also, recrystallization due to the applied strain energy. Hence, at these higher temperatures, some of the microscopic damage that occurred during the low-temperature portion of the thermal cycle may be removed by thermal recovery processes.

To make matters even more complicated, as low-MP alloys, solders typically undergo continual change (i.e., evolution) in their microstructure due to repeated exposure to elevated temperatures. Such changes (grain growth, precipitate particle growth and over-aging, etc.) cause changes in the mechanical properties of the solder, possibly making it less strong, but often making it more brittle, so that it cracks and fractures.

For such a complicated situation, in which temperature changes, thermally induced stresses arise from differential CTEs, properties change, mechanical deformation modes change (from time-independent to time-dependent types), some thermal recovery processes may occur, and microstructure changes, decoupling the problem to consider each effect is far easier than coupling the results to assess the overall situation, as the decoupled parts are *not* independent, as was assumed, to work toward a solution without decoupling contributing factors.

(a)   (b)

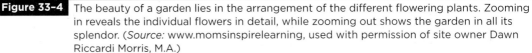

Figure 33-4 The beauty of a garden lies in the arrangement of the different flowering plants. Zooming in reveals the individual flowers in detail, while zooming out shows the garden in all its splendor. (*Source:* www.momsinspirelearning, used with permission of site owner Dawn Riccardi Morris, M.A.)

As a final suggestion when dissecting big and/or complex problems, consider the mental exercise of "zooming in" and "zooming out." By *zooming in*, one is often able to identify and visualize details and events that may involve effects from different stimuli or arising from different phenomena or events. By *zooming out*, one is able to help reassemble the details and integrate the independent solutions for parts of the dissected whole. Zooming out is also extremely important to keeping the overall problem in mind—and in perspective and context.

One never truly knows the beauty of a garden without looking at both the individual flowers and the entire garden (Figure 33–4).

In Summary

Big and/or complex problems are often made less intimidating and easier to solve when they are dissected. How the problem should be dissected can be greatly aided by *zooming in*—to "see the trees"—and *zooming out*—to "see the forest." *Dissecting a problem often means decoupling* multiple phenomena or events to address each individually, as if it were independent of the other phenomena or events, which may or may not be the case. Reassembling the problem involves *coupling*. Coupling is often not so easy, as what were treated as independent phenomena or events may not have been independent but, rather, may have involved partially nullifying or synergistic interactions.

CHAPTER 34

Working Problems Backward and Inverse Problems

THE PATHWAY THAT RUNS from a complex problem to its possible solutions could be likened—at the risk of oversimplification—to the structure of a tree (Figure 34–1). Movement flows from the root of the problem, in its most fundamental and unambiguous formulation, out toward a series of smaller related problems and, from each of these, to a number of potential solutions, some of which are better than others. The roots can be envisioned as feeding material, energy, and information (as in Figure 35–4), as resources or inputs, to the trunk, which represents the central problem to be solved. Different branches could represent prospective pathways to solution of the problem, sometimes leading to other problems to be overcome and, eventually, to viable solutions.

Good decision making in such complex situations, as suggested here, generally requires keeping a number of possibilities open and adopting more than one potential solution. A caveat to be borne in mind is: Whatever choice is made at a decision point—or fork in a branch—pursuing either path (or branch), and solving the smaller problems related to it, results in a large number of other paths—with their related smaller problems and potential solutions—ceasing to be considered. What must be avoided, at all costs, is to have a particular solution take on a life of its own, such that it begins to dictate the perception of the problem rather than it being *an* answer to the problem.

This analogy of a complex problem to the structure of a tree lends credence to both *causal* (e.g., fishbone or Ishikawa) *diagrams* (Chapter 29) and *decision tree diagrams* (Chapter 30).

Figure 34-1 A complex problem is analogous to the structure of a tree, the trunk being the fundamental problem to be addressed, the roots being inputs or resources (e.g., material, energy, information), the larger branches being smaller related problems that arise on an alternative path to solutions, represented by the outermost smaller branches, as viable solutions.

To carry, if not push, this analogy a little further, consider that, after parachuting from an airplane, you landed high in the branches of a tree. To get down from the tree to terra firma, you would work your way back along small branches to forks and larger branches until, finally, you found the trunk of the tree and lowered yourself to safety at its root. The possibilities of starting at the base of a tree and climbing upward and outward toward some goal at the end of one branch or starting out on some branch and trying to find the root of the tree have direct parallels in problem-solving. These are known, generically, as *forward problem-solving* (from root to branch tip—and solution) and *backward problem-solving* (from branch tip to root or basis or input). For problems in science and engineering (as well as in other fields, such as business), the trunk and, particularly, the larger branches could be considered to represent input parameters or model parameters, while the smaller branches could be considered to represent data, outcomes, or solutions. Thus, for a *forward problem:*

$$\text{Model parameters (inputs)} \Rightarrow \text{data (outcomes)}$$

while for a backward problem:

$$\text{Data (outcomes)} \Rightarrow \text{model parameters (inputs)}$$

There are many situations in math, science, and engineering where reverse problem-solving is the most expeditious, if not the only, way to solve a problem. In its simplest form, there is *the problem-solving strategy of working backward.* In its most complex form, there is what is known as *the inverse problem.* Two other backward problem-solving strategies—or reverse approaches—are found in *reverse engineering* (Chapter 16) and in *failure analysis.* The latter involves the systematic examination of failed parts or systems—including pieces, fragments, and debris—to deduce the unambiguous root-cause of the failure. Being a specialized area, rather than a problem-solving technique per se, failure analysis is not considered in this book. Readers interested in the process, however, are encouraged to seek references on the topic (see "Suggested Resources").

In the rest of this chapter, the strategy of working mathematical problems backward and the inverse problem are described and discussed.

Working Math Problems Backward

The strategy of working backward entails starting with the end result(s) and reversing the steps needed to get this (those) result(s) in order to figure out the answer to the problem. Two different types of basic math problems (among more types) that can be solved by this strategy are:

1. When the goal is singular and there are a variety of alternative routes that could be taken. Here, the strategy of working backward allows the optimum route to be ascertained.

    An example is when you are trying to find the best route from your home to some desired destination—say, the town park. You would first look at what neighborhood the town park is in and then trace the best route backward on the map from there to your neighborhood and home.

2. *When end results are given or known in the problem and the initial conditions are sought.*

    An example is when you are trying to figure out how much money you started with if you know how much money you have left and all of the transactions you made during the day.

Working backward to solve a math problem is different from working a problem solved using mathematics in the usual, forward direction. The former is used because it leads to an answer more easily. The latter is used to check oneself (Chapter 6).

Illustrative Example 34-1: Working a Math Problem Backward

If your friend has seven coins whose total value is $0.57, what coins, and how many of each, does she or he have?

Step 1: There must be two pennies, as the option of there being seven pennies would use up all the coins prematurely.

Step 2: You are now faced with finding five coins that total $0.55. Because five dimes would fall short (at $0.50), one coin must be a quarter.

Step 3: Now you need to find four coins that make up the remaining $0.30, and all coins have to be dimes and nickels. The only possibility is to have two dimes and two nickels.

Unlike in engineering school, where inputs to a problem are given and a closed-ended solution is sought, real-world engineering usually involves problem-solving in which the desired end goal or output is known and a viable—if not best—path to that goal or output is sought. Hence, working backward is usually the best—if not only—approach.

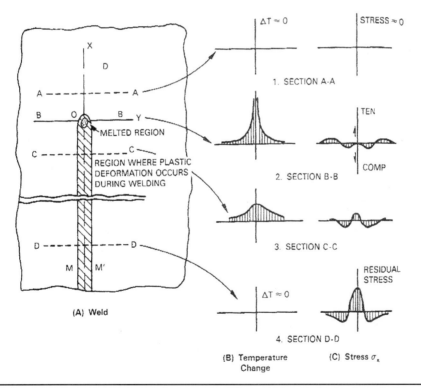

Figure 34-2 Schematic plots of the stresses transverse to the weld line that arise from fusion welding adjacent to the weld pool and behind the weld. These can be estimated by complicated calculations, but they are often measured. In the latter case, the conditions during welding are determined from the measured stresses as an inverse problem. (*Source:* C. L. Jenney and A. O'Brien, editors, *Welding Handbook,* Vol. 1, Welding Technology, 9th edition, American Welding Society, Miami, FL, 2001, p. 306, Figure 7.9; used with permission of the AWS.)

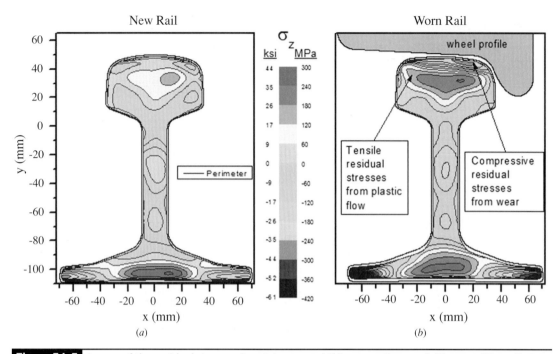

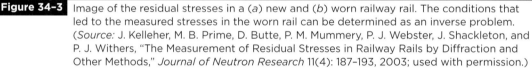

Figure 34-3 Image of the residual stresses in a (*a*) new and (*b*) worn railway rail. The conditions that led to the measured stresses in the worn rail can be determined as an inverse problem. (*Source:* J. Kelleher, M. B. Prime, D. Butte, P. M. Mummery, P. J. Webster, J. Shackleton, and P. J. Withers, "The Measurement of Residual Stresses in Railway Rails by Diffraction and Other Methods," *Journal of Neutron Research* 11(4): 187–193, 2003; used with permission.)

Inverse Problems

An *inverse problem* is a general framework used to convert observed data (e.g., measurements) into information about (as boundary and initial conditions for or inputs to) a physical object or system of interest. An excellent example is if one has measurements of the Earth's gravitational field in a particular geographic area and wishes to find what can be learned about the density of the underlying ground. Solutions to such problems that seek to match field strengths to underlying driving forces or conditions are useful because they provide information about a physical parameter that cannot be directly measured. Thus, inverse problems are one of the most important and well-studied mathematical problems in science and engineering.

Examples where inverse problems arise in science and engineering include, but are not limited to:

■ Stress distributions from processing (e.g., welding; Figure 34–2) or service (e.g., railroad rails; Figure 34–3)
■ Heat transfer problems (Figure 34–4)
■ Remote sensing
■ Astronomy
■ Geophysics and seismology
■ Nondestructive testing (e.g., eddy current measurements)
■ Medical imaging
■ Computer vision

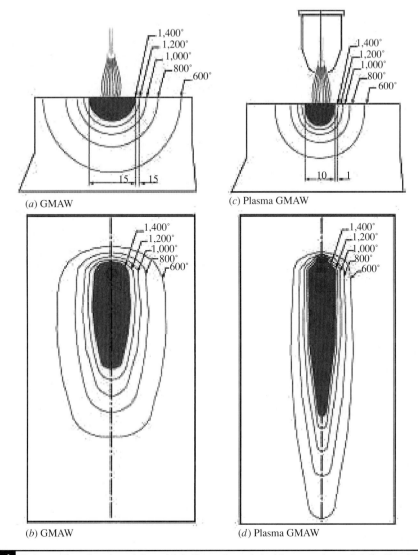

(a) GMAW

(c) Plasma GMAW

(b) GMAW

(d) Plasma GMAW

Figure 34-4 Schematics showing: (a) and (c), the extrapolated temperature distribution in a transverse direction with respect to the weld line (with the depths of penetration normalized); (b) and (d), illustration of the observed high-temperature isotherm distribution around the moving heat sources. (*Source:* Igor Dykhno and Raymond Davis, "Joining GMAW and GTAW," *The Fabricator,* November 7, 2006, Figure 9; used with permission.)

Since the mathematics involved in inverse problems is advanced and complicated (e.g., partial differential equations for which numerical solution is required; see Chapter 13), no specific illustrative example will be given.[1] Rather, an example of how the technique can be used will be given.

---

[1]The reason inverse problems, and working backward, were not covered in Part 1 on mathematical problem-solving techniques but, instead, in Part 4 on conceptual techniques, is that the conceptual strategy is deemed more important than the specific mathematics.

Figure 34-5 Photograph of the surface of Jupiter showing its patterns of weather. Condition on the surface of the planet can be estimated by solving the inverse problem. (*Source:* NASA's *Planetary Journal,* with full credit but no requirement for permission.)

Illustrative Example 34-2: An Example of an Inverse Problem

A very common area for the application of inverse problems is heat transfer. Whether the transfer process is dominated by conduction, convection, or radiation, or some combination of these, it is often important to work backward toward the boundary conditions and input parameters that gave rise to a measured pattern or distribution of temperature in a 2D or 3D body. Such a technique is used to assess what is happening on Earth to cause ocean currents and/or weather patterns, as well as on distant planets, moons, and stars (e.g., our Sun) to understand dynamic processes or events that may be going on under their surfaces (Figure 34-5).

Table 34–1 gives several examples of where inverse problems are used.

TABLE 34-1 Summary of Areas for Inverse Problems		
Application Area	**Data Outcome/Measurements**	**Model Parameters/Inputs**
Heat Transfer	Temperature distribution field	Weld/casting geometry Material thermal properties Welding/casting parameters Defects
Stress Analysis	Stress or strain field	Part geometry Material properties Shrinkage/cracking Distortion
Seismology	Seismic data	Earthquake source Earthquake magnitude
Medical Imaging NDT	Signal strength distribution	Internal structure Defects Sizing/dimensioning
Astronomy	Light intensity Color/Red-shift	Astronomical bodies Black holes

In Summary

Working backward has value in the solution of some problems. A strategy of working even seemingly simple math problems backward is often the most expeditious method for finding an answer. For problems in which there are data in the form of measurements (temperature, gravity, stress, etc.), the mathematical approach of *inverse problems* is often essential. As we learned earlier, working backward is how designers learn from prior designs in *reverse engineering* or from failures in *failure analysis.*

Suggested Resources

Berk, Joseph, *Systems Failure Analysis,* ASM International, Novelty, OH, 2009.

Kirsch, Andreas, *An Introduction to the Mathematical Theory of Inverse Problems,* 2nd edition, Applied Mathematics Series 120, Springer, Dordrecht, The Netherlands, 2011.

Martin, Perry L., *Electronic Failure Analysis Handbook,* McGraw-Hill Handbooks, New York, 1999.

McEvily, Arthur T., *Metal Failures: Mechanisms, Analysis, Prevention,* John Wiley & Sons, Hoboken, NJ, 2002.

Young, Gregg, *Reasoning Backwards: Sherlock Holmes' Guide to Effective Problem Solving,* Young Associates, Inc., Midland, MI, 2011.

CHAPTER 35

Functional Analysis and Black Boxes

WHEN ENGINEERS HAVE to deal with a system comprising many component subsystems, component parts, processes, or operations, usually as part of the design process for a new product or process, they will typically employ the techniques of *functional analysis* and *black boxes*. In its most basic form, *functional analysis* (also known as *function analysis,* except that this latter term can easily be confused with the analysis of functions in a branch of mathematics) considers the activities or actions that must be performed by each subsystem and component, process or subprocess, and/or operation in order to achieve the desired outcome at each level, step, or stage, and, in turn, with proper integration, in the overall system. Simply put: Functional analysis identifies the transformations necessary to turn available inputs of materials, energy, and/or information into the desired output as material(s), energy, and/or information.

A *black box* in science and engineering is a device, object, or system that can be viewed solely in terms of inputs to it and outputs from it by its transfer characteristics (as details) without any knowledge of the internal workings (i.e., details) of the box. The implementation of a black box is "opaque," as opposed to the implementation of a "transparent" (clear) box.

These two problem-solving techniques are treated together in this chapter because they are closely related. Black boxes are frequently what are identified, as opposed to exact devices and the like, during functional analysis.

Functional Analysis

A complex *technical system* (e.g., an automobile, an airplane, or a computer) is commonly broken down—or decomposed (see Chapter 33)—into assemblies (or subsystems) and components within each assembly (or subsystem). (See Figure 35–1.)

Illustrative Example 35–1: Breakdown of a Technical System

A great example of a complex technical system that is commonly broken down into subassemblies and, within subassemblies, into components, is an airplane—say, a commercial airliner (Figure 35-2).

At its most basic, a commercial airliner (or transport aircraft) is built up from four major subassemblies: fuselage (subassembly 1); wings, including control surfaces (subassembly 2); engines/nacelles (subassembly 3); and empennage (subassembly 4).[1]

---

[1] The airframe builder (e.g., Boeing) designs and builds the nacelles, with intakes, mounting structure, and exhausts, while an engine manufacturer (e.g., General Electric) designs and builds the actual engines (e.g., gas turbines) under a contract from the airframe company. The aircraft manufacturer (e.g., Boeing) provides specifications for needed engine performance and installs delivered engines.

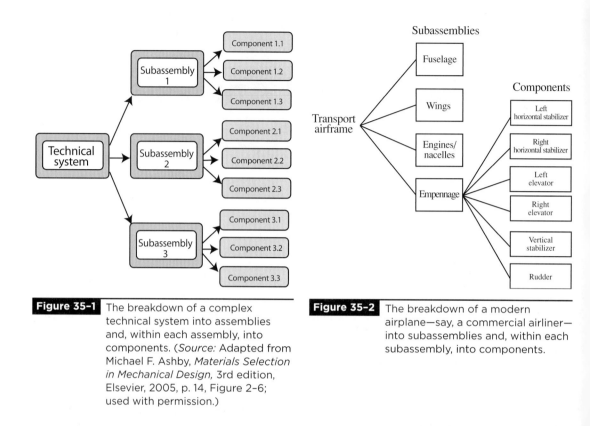

Figure 35-1 The breakdown of a complex technical system into assemblies and, within each assembly, into components. (*Source:* Adapted from Michael F. Ashby, *Materials Selection in Mechanical Design,* 3rd edition, Elsevier, 2005, p. 14, Figure 2–6; used with permission.)

Figure 35-2 The breakdown of a modern airplane—say, a commercial airliner—into subassemblies and, within each subassembly, into components.

So as not to become burdensome, because of the tremendous complexity of such an aircraft, only the empennage assembly (Figure 35–3) is further broken down into components. These components include left and right horizontal stabilizers (components 3.1 and 3.2)[2], left and right elevators (components 3.3 and 3.4), vertical stabilizer (component 3.5), and rudder (component 3.6), all of which consist of internal stiffeners, as well as of skins.

For structures as complex as many of the components of a major assembly for an airplane (e.g., a horizontal stabilizer), there can be dozens to hundreds of parts. At some point in the design of an airplane, the simple breakdown of the technical system shown in Figure 35–2 has to be broken down further—for example, making the horizontal stabilizer a subassembly and identifying major components of it (upper and lower skins or "covers," spanwise beams or spars, fore-to-aft ribs, leading-edge closure fairing, etc.). Obviously, there are many more small parts (shear-transfer clips, linkages, etc.) as well as hydraulic lines, valves and actuators, and/or wires and servomotors for controls

This manner of decomposing a complex technical system is useful for analyzing an existing product but tends to be more restrictive (i.e., not detailed enough) to allow a new product to be designed (i.e., for synthesizing a new design).

A better approach for decomposing a complex product to allow design is one based on the ideas of *systems analysis,* using the technique of functional analysis. *Functional analysis* creates an arrangement

---

[2]Because of their large size, the horizontal stabilizers for transport aircraft (commercial and military) are rarely one piece but, rather, consist of largely mirror-image left and right units. In smaller aircraft, this assembly could be one piece. The same is true for the wings of most airplanes, even the smallest ones.

Figure 35-3 Photograph showing the empennage assembly for an airplane-here, a Boeing 747–200 commercial airliner. (*Source:* Wikimedia Commons; original photograph by Dtom on 9 June 2008, released to the public domain by the photographer.)

known as a *function structure* or *function decomposition* of the system. In functional analysis, the system is decomposed into subsystems for which the designer thinks about the inputs, flows, and outputs, with the details within the subsystem converting material, energy, and/or information into the desired outputs of material, energy, and/or information from that subsystem as the inputs to the next linked subsystem (Figure 35–4). This continues until the needed/desired outputs of material(s), energy, and/or information is obtained from the overall integrated system.

The function structure obtained from functional analysis is abstract, in the sense that there usually are not—or need not be—details about each subsystem function. All that is trying to be accomplished at this stage of design is to better understand what is needed in the system at a fairly high level, thereby allowing such abstraction. Because the subsystems in a function structure are abstract, they are often

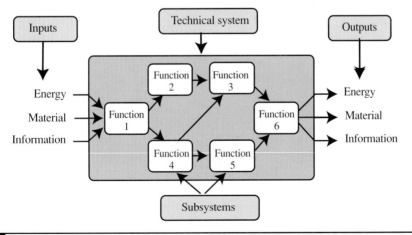

Figure 35-4 The breakdown of a complex system into key functions within systems engineering using functional analysis to create a function structure. (*Source:* Adapted from Michael F. Ashby, *Materials Selection in Mechanical Design,* 3rd edition, Elsevier, 2005, p. 14, Figure 2-7; used with permission.)

treated as *black boxes*. Once again, the black box is simply a representation of what hardware (for example) will be needed to transform given inputs into needed outputs, without any knowledge of the integral workings of the box.[3]

The use of black boxes in this abstract way is quite common in electronics and in electromechanical and/or "mechatronic" systems. In electronics, such a black box can be a sealed piece of replaceable equipment, known as a *line-replaceable unit* (LRU). In computers, "black box" often refers to a piece of equipment provided by a vendor. The vendor often supports such equipment, with the company receiving the black box typically remaining hands-off. Of course, the infamous black box—which is actually painted bright international orange for visibility—carried on all airplanes, commercial and military alike, contains instrumentation for recording voice communications to and from the aircraft, along with data-recording instruments that monitor critical systems (hydraulics, electrical power, engines, environmental systems, etc.).

In Summary

Functional analysis allows complex systems to be broken down or decomposed into key functions with a function structure. Being abstract at the beginning of the design process, such systems analysis and function structures can be and usually are abstract, showing only inputs of material, energy, and information that lead to outputs of material, energy, and information after transformation within a *black box*.

---

[3]The opposite of a black box is a subsystem (or, more appropriately, a function in the function structure) wherein the inner components (or logic) are available for examination in what is known as a "white box," "glass box," or "clear box."

CHAPTER 36

Morphological Analysis and Morphological Charts

MORPHOLOGICAL CHARTS ARE another in a series of charts, matrices, diagrams, or tables that represent planar spaces in which visual elements of a design or problem-solving space can be used to advantage. While the end product of a morphological chart is almost always graphic or iconic—which then guides pathways to a problem solution—the initial approach is definitely abstract, as it occurs during conceptualization. *Morphological analysis,* which precedes the creation of a morphological chart, is a method for generating ideas in a systematic and analytic manner. Morphological analysis, and the resulting chart, is most often used at the conceptual design stage of the design process. For this reason, morphological charts are often listed as one of the ways to stimulate thinking and elicit creativity.[1]

As discussed in the previous chapter (Chapter 35), a starting point for designing a new product is to consider the required functions. The various functions and subfunctions of a product and its component parts can be established through *functional analysis* (or *function analysis*). But functional analysis surely does not guarantee that all relevant functions or subfunctions have been identified, or that those that have been identified are unique. It is not at all uncommon that a number of approaches or solutions to some, if not most or all, functions and subfunctions are already known by a designer, perhaps from prior experience or from reverse engineering (Chapter 16). However, other, possibly innovative or revolutionary, approaches or solutions may come out of the designer's mind. Together, these known and new approaches or solutions form the cells in the morphological chart. The morphological method yields a matrix of solutions to provide needed functions.

The word *morphology* refers to the study of shape or form (from the Greek *morphe,* "form"). Thus, *morphological analysis* is a way of creating new forms. The intent of morphological analysis during the conceptual stage of design is to bring order out of an abstract or, at least, fuzzy problem solution. In the process of such analysis, factors that might not ordinarily emerge otherwise may be seen to be able to be combined. Morphological analysis is a systematic and structured approach to more fully and clearly define a problem and possible solutions that employ a simple matrix in the form of a chart or box.

Morphological analysis works best when the problem to be solved can be decomposed into subproblems or components, for example, using functional analysis (Chapter 35). Each proposed subproblem or component should represent an identifiable and meaningful part of the overall problem.

A *morphological chart* is a table or 2D matrix that is based on a prior functional analysis. Down columns, at the left-hand side of the table or chart, the required functions (from the functional structure) are listed. Across the top, from left to right, are various options to achieve each function, row by row. In boxes or cells across the chart, simple sketches are typically used, although concise words could also be used. The use of simple sketches tends to serve as a visual aid and stimulates other, different ideas.

---

[1]There are some (albeit a minority) who feel morphological charts are *not* for idea generation but, rather, to organize and synthesize solutions from partial solutions. Morphological charts can certainly be used this way, but they are absolutely a useful way to generate ideas!

Figure 36–1 is a nice example (found on www.eng.fsu.edu/~haik/design/idea_generation.html) of a morphological chart for methods of harvesting vegetables that grow under the ground, that is, root vegetables. The key functions that emerged from functional analysis are, in order:

1. Vegetable picking device
2. Vegetable placing device
3. Dirt sifting device
4. Packaging device
5. Method for transportation
6. Power source

Simple sketches appear in the boxes of the chart representing options by which the various functions could—conceptually—be achieved.

Once the morphological chart is filled in, it is advisable to generate several feasible designs for the overall system in which different mechanisms are used for the various functions. The best among these alternative designs can be found by building and testing models or prototypes (see Chapters 20 and 22) or by market analysis, for consumer-driven products.

For this nice example from Florida State University, interested readers are encouraged to visit the aforementioned website to see various concepts that emerged, one of which is shown in Figure 36–2 in which the following options were employed:

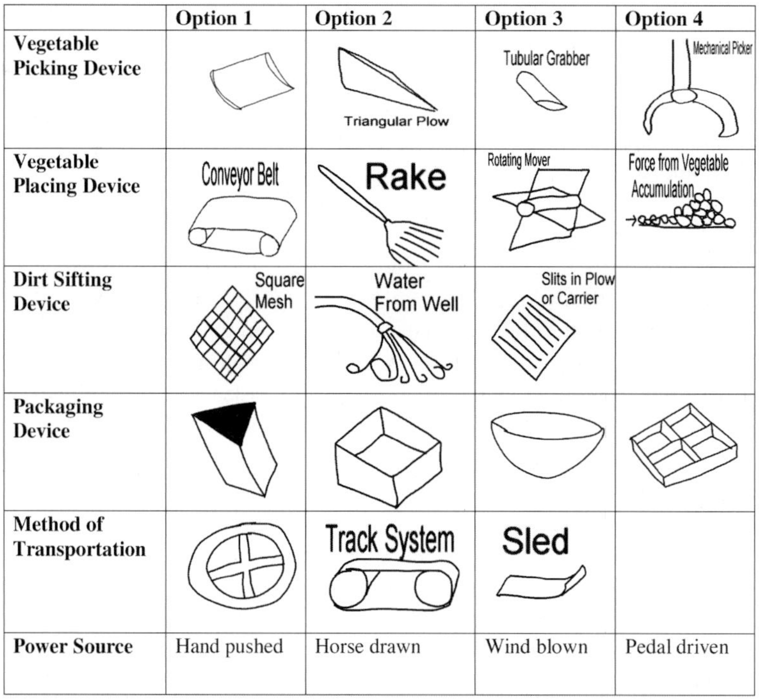

	Option 1	Option 2	Option 3	Option 4
Vegetable Picking Device		Triangular Plow	Tubular Grabber	Mechanical Picker
Vegetable Placing Device	Conveyor Belt	Rake	Rotating Mover	Force from Vegetable Accumulation
Dirt Sifting Device	Square Mesh	Water From Well	Slits in Plow or Carrier	
Packaging Device				
Method of Transportation		Track System	Sled	
Power Source	Hand pushed	Horse drawn	Wind blown	Pedal driven

Figure 36-1 A morphological chart of a harvesting machine for root vegetables depicting various options (across the top of the chart) to achieve various functions listed vertically down the left side of the chart. The various functions come from prior functional analysis, while the possible solution options can and frequently do come from group or individual brainstorming. (*Source:* Professor Yousef Al-Hayek, formerly of Florida A&M/Florida State University; used with permission.)

Figure 36-2 One possible design concept (Concept 2 on the website) for harvesting root vegetables incorporating a steel-edged plow to pick the vegetables from just beneath the ground, a conveyor belt for sifting out dirt while placing vegetables onto a guide from which the human operator moves them into crates. The machine is drawn by a horse. (*Source:* Profesor Yousef Al-Hayek, formerly of Florida A&M/Florida State University; used with permission.)

Vegetable picking device = Option 1
Vegetable placing device = Option 3
Dirt sifting device = Option 1
Packaging device = Option 2
Method of transportation = Option 1
Power source = Option 2

The idea is said to stem from modern potato harvesting equipment. A steel plow (chosen to resist wear from the abrasive soil) is able to rip potatoes—which grow close to the surface of soil—free from roots, whereupon they move up to a plow and onto a hemp conveyor belt. Hemp was chosen, as it is an extremely strong natural fiber. The design of the conveyor system sorts vegetables from the dirt and small stones, with the vegetables staying on the belt and solids and debris falling through. The conveyor is driven by the rear wheels via a 3:1 gear ratio, allowing the belt to move three times the speed of the tangential velocity of the rear wheels. Once at the top of the conveyor system, vegetables are pushed into a rear-mounted loading table complete with guides. The operator controls the horses and directs vegetables into a basket attached to the rear of the machine.

Illustrative Example 36-1: Using a Morphological Chart

Professor Joshua D. Summers, of Clemson University's Mechanical Engineering Department, gave an interesting example of the use of a morphological chart on a former version of his website at http://www.Clemson.edu/ . . . /Images/d/d2/14 -MorphologicalCharts.pdf for a human-powered golf ball retriever.
    Functional analysis of the problem identified the following functions as being required:

- Find hole.
- Move to hole.
- Grab ball.
- Control ball.
- Convert power.

- Release ball.
- Guide ball.
- Store ball.

The resulting chart is shown in Figure 36–3, in which, for example, four possible means for finding the hole are given as (1) precalculated distance (e.g., using GPS data), (2) sonar, (3) tactile, and (4) sight. Other function rows were similarly populated with optional means (albeit as words, not simple sketches or icons).

Three possible solution combinations (from among a total of $4^8 = 65,536$ possible combinations) are identified by three paths (shown in Figure 36–3). Choosing one solution over another would probably require testing of a model or prototype.

In lieu of morphological charts for two-dimensional situations (i.e., a variety of options for a list of functions), *morphological boxes* can be created. These are best illustrated with an example.

Illustrative Example 36–2: Using Morphological Boxes

In George E. Dieter's classic textbook *Engineering Design: A Materials and Processing Approach,* he gives an excellent example drawn by him from J. R. M. Alger and C. V. Hays, *Creative Synthesis in Design.* The example (used here with permission) involves the development of an innovative design for an automatic electric dryer.

As the first step, three major functions that must be performed by a clothes dryer were identified as:

1. *The source of heat,* which determines the speed and uniformity of drying
2. *The environment around the clothes,* which determines the smell of the dried clothes
3. *The drying mechanism,* which determines the degree of wrinkling of the dried clothes

Figure 36–4 shows the various possible ways for accomplishing each of the three dryer functions listed here. Each cell or box in the 3D array represents a particular drying mechanism, heat source, and environment, yielding $5 \times 4 \times 4 = 80$ potential design concepts. Some combinations can easily be eliminated, as they are impractical. Others will merit further study, possibly involving testing, prototyping, and/or market analysis.

Dieter makes this essential point: "The chief virtue of the morphological box is that it provides a formal mechanism by which unusual or innovative problem solutions will not be overlooked in the thought process."

Find Hole	Pre-calculated	Sonar	Tactile	Sight
Move to Hole	Track	Sail	Float	Telescope
Grab Ball	Pinchers	Suction	Adhesive	Pierce
Contain Ball	Cup	Bag	Net	Tube
Convert Power	Gears	Linkage (crank)	Pump	Ratchet
Release Ball	Poke	Jiggle	Slice	Heat
Guide Ball	Track	String	Magnetic	Fingers
Store Ball	Bucket	Cartridge	Hole	Bag

Figure 36-3   A morphological chart for the design of a human-powered golf ball retriever. Three designs (from among $\mathcal{A}8 = 65,536$ possibilities) are shown. (*Source:* Joshua D. Summers, assistant professor in the Mechanical Engineering Department at Clemson University; used with his kind permission.)

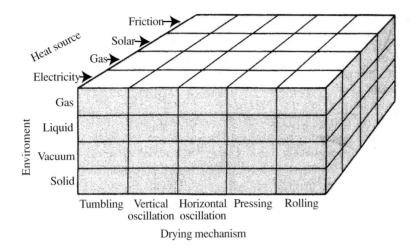

Figure 36-4 A morphological box for the design of an innovative automatic clothes dryer. The three key functions of a dryer—heat source, environment for the clothes during drying, and drying mechanism—obtained from prior functional analysis, are used as axes. Those boxes representing "best designs" from testing or market analysis are typically subjected to further morphological analysis (see Figure 36-5). (*Source:* From George E. Dieter, *Engineering Design: A Materials and Processing Approach,* 2nd edition, McGraw-Hill, New York, 1991, Figure 3, p. 130, taken by him from J. R. M. Alger and C. V. Hays, *Creative Synthesis in Design,* Prentice-Hall, Englewood Cliffs, NJ, 1964; used with permission.)

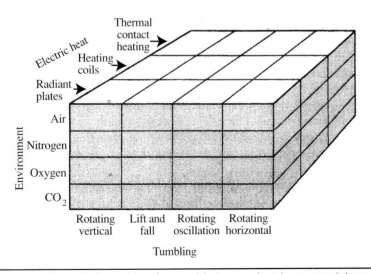

Figure 36-5 A second morphological box for considering an electric-powered dryer using a tumbling mechanism to dry clothes in a gas environment shows the many options for each axis. (*Source:* From George E. Dieter, *Engineering Design: A Materials and Processing Approach,* 2nd edition, McGraw-Hill, 1991, Figure 3-7, p. 131, taken by him from J. R. M. Alger and C. V. Hays, *Creative Synthesis in Design,* Prentice-Hall, Englewood Cliffs, NJ, 1964; used with permission.)

Dieter's example goes on to repeat the morphological analysis for the most promising cells, one of which is shown (Figure 36–5) for the usual design solution involving an electric heat source, a tumbling drying action, and a gas environment surrounding the clothes.

In Summary

Morphological analysis considers the shape or form of a new product during the creative conceptual design stage. Typically, a 2D *morphological chart* is constructed with various options (shown across rows) for meeting individual functions obtained from functional analysis (listed vertically). Simple sketches or icons in each cell tend to be used to allow quick and easy grasping of the idea. For more complicated situations, a 3D *morphological box* can be used.

Suggested Resources

Alger, J. R. M., and C. V. Hays, *Creative Synthesis in Design,* Prentice-Hall, Englewood Cliffs, NJ, 1964.

Dieter, George E., *Engineering Design: A Materials and Processing Approach,* 2nd edition, McGraw-Hill, New York, 1991, pp. 130–131.

CHAPTER 37

Storyboarding

STORYBOARDS ARE GRAPHICAL ORGANIZERS, predominately used for the purpose of previsualizing an artistic creation, such as a motion picture, animation, motion graphics, an interactive media sequence, or even the storyline of a novel. Few would be surprised to learn that the *storyboarding process* was developed at the Walt Disney Studios during the early 1930s. Similar processes had been used since the 1920s at both Walt Disney and other animation studios. As opposed to full storyboards, these earlier techniques were "story sketches" with a comic book look.

Whatever their origin or no matter what they are called, there are those who would question whether the technique has any role in engineering problem-solving. To be honest, the author also questioned himself whether the technique should be included in the collection of techniques covered in this book. In the end, after recalling a couple of instances when the path of a technical project was planned using storyboarding, the answer was a tepid "yes." So here goes.

It was rumored that the brilliant, successful, but highly eccentric entrepreneur Howard Hughes adopted and adapted storyboards he had seen used in the film industry (in which he had business and financial interests) to business outside the creative arts, supposedly at Hughes Aircraft.

True or not, *storyboards* and *storyboarding are* used today by industry for planning advertising campaigns, proposals, and business presentations intended to influence people to action. Without question, there are large engineering projects that can benefit greatly from "design comics" or storyboards to logically structure a problem-solving strategy using simple graphic images sequenced to show steps and flow from start to desired finish. An example was the use of storyboards at Grumman Aerospace Corporation during planning of the mission to land a man on the Moon with the design of the Lunar Excursion Module (LEM).

An advantage of *storyboarding* is that it allows the user to experiment with changes in the "storyline" or plan, either to create a more effective solution or to elicit greater, more favorable response. Another advantage of storyboarding is that it is a process of *visual thinking* and planning that allows a group of like-interested people to brainstorm together (see Chapter 31), placing their ideas on storyboards and arranging the storyboards on the wall with the goal of developing a logical flow or "story." Likewise, the technique of visual thinking can be extremely useful for an individual, working alone, to capture, display, and organize ideas as they emerge. In either case, acting as a group or as an individual, Post-its bearing simple sketches work very well for this process.

In short, the process of storyboarding and the use of storyboards facilitate rapid expression of ideas that form in one's mind (or in the minds of several people involved in group brainstorming). Images often elicit new ideas, as other images, but, even if they do not, storyboards bearing images can be easily rearranged to explore alternative sequences of events or actions to create alternative pathways to a problem solution or other goal. Visualization is extremely powerful—even athletes use it to "see" themselves succeed in order to succeed.

Most exciting of all is the possibility of using storyboarding on those wonderfully fulfilling instances when engineering takes a foray into creativity, and innovative and even life-altering ideas emerge.

Figure 37–1 shows a typical storyboard.

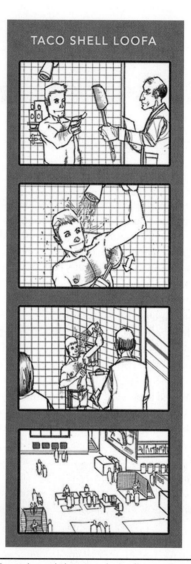

Figure 37-1 An example of storyboards and the storyboarding process widely used in the creative arts, into which engineering can move on occasions, is shown: a storyboard for a prospective TV ad campaign for Taco Bell. The idea was for Taco Bell scientists to bring college student subjects into their lab to study further uses of taco shells, here as a "Taco Shell Loofa." (*Source:* Wikimedia Commons, under an image search, originally contributed by N8VanDyke on 9 February 2010, and used under free Creative Commons Attribution 3.0 Unported license.)

In Summary

Storyboarding, so important to the creative arts (film, theater, novel writing, etc.), can have value to engineers in problem-solving. It can offer particular advantage when creativity is required to find a solution. The benefit is usually to force a plan and allow plan alteration before action is taken and serious money is spent. The saying goes: Plan your work, and work your plan!

CHAPTER 38

Iteration and the Iterative Design Process

ITERATION IS DEFINED as the act of repeating a process (or step in a process), over and over again, with the aim of approaching (or converging on) a desired goal, target, or result (e.g., problem solution). Each repetition of the process is called an *iteration*. The results of each progressive iteration are used as the starting point for the next iteration. In mathematics and computing, the results of one iterative step are used as actual inputs to the next iterative step. In design, or other problem-solving for which the solution is not a number, the results of one iterative step are used to guide some change in the outcome of the next iterative step, often using guided empiricism within a trial-and-error approach (see Chapter 23).

    Iteration in mathematics can apply to either of two general situations. The first situation is to apply a function repeatedly using the output from one iteration as the input to the next iteration. A familiar and fascinating application of the iteration of a mathematical equation or function is in chaos theory and the new discipline of fractal geometry—for example, the Mandelbrot equation $Z_{n+1} = Z_n^2 + C$ or, more generally, $Z_{n+1} = Z_n^m + C$. The result is the creation of *self-similarity* (see Chapter 8) at ever-smaller scales (Figure 38–1), known as a *Mandelbrot set*. The excitement over fractal geometry is

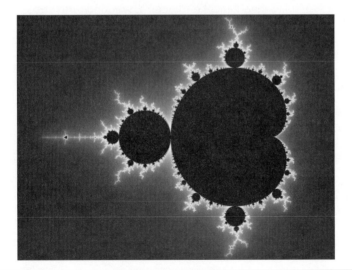

Figure 38-1   The fascinating plot of the Mandelbrot equation, $Z_{n+1} = Z_n^2 + C$, for which the same pattern repeats and repeats at ever finer scales—or on larger and larger scales—ad infinitum when the output of the equation for one input value of Z is used as the input for the next iteration of the equation. (*Source:* Wikimedia Commons, under an image search; originally contributed by Wolfgangberger on 4 December 2006, and used under free Creative Commons Attribution Share Alike 3.0 license.)

that, as opposed to euclidean geometry that deals with smoothness, typical of works by human beings (especially engineers and architects), fractal geometry deals with roughness, as found in Nature.

The second situation for iteration in mathematics is as a method to produce approximate numerical solutions to certain mathematically based problems (see Chapter 13).

In computing, *iteration* involves the repetition of a process or step in a computer program. Such an operation is known as a *recursive* step or operation.

In structural analysis during design, *iteration* is often necessary for complex structures that are to be optimized for some factor, such as to minimize weight, when changing one factor in the design changes some other factor in the design and the analysis.

An illustrative example should help.

Illustrative Example 38–1: Iteration to Minimize the Weight of an Airframe

Weight is critical in aircraft and spacecraft, as reduced weight results in increased performance (e.g., reduced fuel consumption, greater maneuverability, and/or greater speed) and/or greater fuel capacity and/or payload capacity. This is particularly true in military aircraft, for example.

But minimizing the weight of the airframe for a fighter, for example, is nontrivial and must be done iteratively. The reason is: How do you reduce the weight of any structural element to its practical limit *before* the airframe and, thus, its elements have been designed?

The process of designing the airframe for a fighter begins with a specified mission, which, in turn, places requirements on the aircraft's size, speed, maneuverability, range, and so on, and, in turn, its structural components (wings, horizontal and vertical stabilizers, etc.).

The aeronautical engineer (or aerodynamicist) develops a suitable vehicle configuration, as fuselage, wing, and control surface sizes and shapes, to achieve the desired mission. Structural designers (usually mechanical engineers), working with stress analysts (often civil engineers to handle the complex truss structures) then design a structural framework (e.g., airframe) for this vehicle capable of tolerating the loads that will be applied by the mission requirements. But, besides externally applied loads from g-forces, wind, drag, and so on, there will be internal loads from the mass of each component and their effects on one another (i.e., inertia loads from component mass—and weight). Applied loads (e.g., F) can be reasonably estimated from the mission requirement (e.g., acceleration a), but not completely or precisely, as the mass m cannot be determined accurately until the structural components have been designed.

You see the problem?

In short, it goes like this:

■ Rough out a basic structure (e.g., the wing) to provide the required aerodynamics and then rough out structural details (upper and lower skins, spanwise and chordwise stiffeners, ribs, etc.) that will tolerate the estimated applied loads and some nominal internal loads.

■ Estimate the weights of all of the components.

■ Generate the internal loads that will arise from components of these sizes—and their weights.

■ Add the internal loads to the applied loads, adjusting both types of loads as necessary to reflect the new, more detailed structure and correct for mass effects.

■ Recalculate stresses.

■ Refine the design to reduce weight (by reducing size—especially cross-sectional area) of all components.

- Recalculate weights.
- Regenerate internal and adjusted applied (external) loads.
- Repeat until a point of diminishing returns is reached.[1]

To make the problem even tougher, consider this. Every reduction in weight for the airframe allows the power plant (e.g., jet engine) to be reduced in thrust capability and, thus, size and weight. A lighter engine, with less thrust, requires a less substantial airframe to support it. A less substantial airframe weighs less, so an even smaller, lower-thrust, lighter-weight engine would work. Etc. Etc. Etc.

Just for completeness, even though it is only peripherally related to engineering per se, via the design process, explanation of *iteration* as applied to project management will be left simply as: the technique of developing and delivering incremental improvements of a product (during its development) or a process (during its optimization) using guided empiricism.

The Iterative Design Process

Iterative design is a methodology based on a cyclic process of modeling, prototyping, testing, analyzing, and refining a product or a process (see Chapters 18 through 22). The goal of the cyclic process is to ultimately improve the functionality and quality of a design. In later stages, including during early production, an iterative process may be used to reduce costs in what is known as *producibility.*

While the iterative process may be applied throughout the development of a new product or process, it is most expeditiously applied in the earliest stages (e.g., possibly during conceptual design to identify alternatives, but, more often, during embodiment design or tradeoff studies, where decision making leads to a preferred approach or approaches. Changes at early stages are easier to make and are less expensive than changes made later.[2]

Evaluation of any prototype should involve a *focus group.* Such a group should, obviously, include those with a direct vested interest in the project, as well as intimate knowledge of the prototype (e.g., the designer), and some others who are not directly associated with the product or process in order to secure totally nonbiased opinions. Information obtained from the focus group should be analyzed (synthesized) and incorporated (at least to some degree) into the next iteration of the design. The process should be repeated until user issues have been reduced to an acceptable level.

Figure 38–2 shows a nice schematic of the iterative design process cycle.

Illustrative Example 38–2: The "Marshmallow Challenge"

The so-called Marshmallow Challenge is an instructive design challenge with profound implications.[3] It involves the seemingly simple task of constructing the tallest possible freestanding structure with a marshmallow top. The structure must be completed by a team within 18 minutes using 20 sticks of spaghetti, a yard of tape, and one yard of string— and a marshmallow.

Observations and studies of various teams have repeatedly shown that kindergartners are regularly able to build a taller structure than groups of business school graduates and, sadly, engineers. This result is explained by the tendency for children to stick the

---

[1]"Diminishing returns" are reached when any further reduction in weight (in this example) drives cost up at a rate considered inordinate for the gain in weight reduction and associated performance improvement(s).
[2]Late changes are often (but not always) the result of less-than-thorough conceptualization and tradeoff studies.
[3]The challenge was invented by Peter Skillman of Palm, Inc., and was popularized by Tom Wujec of Autodesk.

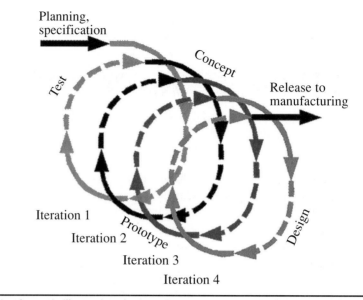

Planning, specification

Test

Concept

Release to manufacturing

Iteration 1

Iteration 2

Prototype

Iteration 3

Design

Iteration 4

Figure 38-2 A schematic illustrating the iterative process as applied to product design, from specification of needs to release of information (drawings, procedures, etc.) to manufacturing.

marshmallow on top of a very simple structure, test the prototype, and continue to improve upon it. Business school graduates—and some engineers—on the other hand, tend to spend inordinate time vying for power, planning, and analyzing, and finally producing a structure to which the marshmallow is added—with little time for iteration.

In Summary

Iteration is the process of running through a cycle of planning, concept generation, preliminary design, prototyping, testing, and repeating the cycle again and again until the desired goal is reached. Iteration is used in mathematics, computing, and problem-solving, including design. In mathematics, the output of one iterative step is the input to the next iteration. In design, the outcome of one iterative step is used to empirically guide changes for the next iteration.

Remember the Marshmallow Challenge, in which the lesson is to try something—not to the exclusion of thinking, but to guide or direct thinking with practical knowledge.

Closing Thoughts

As ONE WHO proudly considers himself an engineer who teaches, I may be somewhat biased, but I would be willing to debate with anyone—including a medical doctor or a lawyer—that there is no profession more important, and perhaps noble, than engineering. At the risk of oversimplification: Decisions made by medical doctors usually affect one person at a time, so *one* might die due to a mistake. Lawyers generally defend one person at a time, so failure would put *one* innocent person in jail. But engineers design and oversee the building of things that affect tens, hundreds, thousands, tens of thousands, or more at a time—buses, planes, bridges, skyscrapers, nuclear power plants, and so on. When an engineer fails, the consequences can be horrific. And, while doctors are always known by their patients and lawyers are always known by their clients (and, in a modern media-driven world, both may be known by thousands, tens of thousands, or millions of readers, listeners, or viewers of Dr. Oz or Judge Judy), engineers are seldom known by those who enjoy engineering successes but might be headline news due to a serious engineering failure. From this standpoint, engineers are also noble, doing what they do without fanfare.

Engineers solve problems to benefit society (Chapter 1). To solve problems effectively and efficiently, engineers use a variety of problem-solving techniques. Regrettably, when young people, desiring to become engineers, go to engineering school, they are seldom, if ever, taught *techniques* for solving problems. Rather, they are given huge numbers of problems to solve in the hope, if not expectation, that by succeeding in solving a problem, they will have learned a problem-solving technique. This is *not* necessarily true! Few really think about what they did to succeed. They solve the problem, but, relieved to have done so, may not be consciously aware of the technique they used to solve it. If they are not aware, or are not made aware, they have *not* learned a problem-solving technique. They simply solved *a* problem.

Imagine if a baseball player as a batter trying to hit a pitched ball had no idea what he had to do to hit the ball or, if he did hit the ball, what it was that he did to do so! Few would succeed at the game. Ergo, well less than half of those who enter college to become engineers graduate with a degree in engineering. Some flunk out because, like a person being thrown in the deep end of a pool to learn to swim, they drown in the process. Others, in struggling to survive, with no idea how to succeed, get so upset by the anxiety of the experience, choose to leave the pool, and never swim again. Of course, some manage to learn to swim, and, as with riding a bike, once they figured out what they needed to do, they were able to repeat it.

Once engineering students graduate with a bachelor's degree in engineering, they really begin to learn engineering. Increasingly in modern, research-driven universities, engineering faculty teach engineering courses but have not worked—and earned a living—as an engineer. Therefore, it would be serendipitous[1] if they truly knew how a practicing engineer solves problems. Imagine wanting to learn to play the piano. Would you not want to go to someone, as a teacher, who knew how to play the piano, as opposed to one who just read—and lectured—a lot about how pianos work?

Practicing engineers solve problems. Students of engineering really learn to solve problems by

---

[1] *Serendipitous* means "as the result of happenstance and good fortune."

working with experienced engineers. Those who are observant, or who humble themselves to ask, learn the host of techniques by which problems can be solved from the "masters."

Look at the front cover of this book if you want to see the work of masters. The Great Pyramids of Giza were built during the Fourth Dynasty between 2580 and 2560 BC, the greatest of all to honor the beloved Pharaoh Khufu (better known as Cheops). With a height (i.e., the pyramid's altitude) of 146 m (480.69 ft), Khufu's pyramid was the tallest structure built by humans for more than 3800 years. Oldest of the Seven Wonders of the Ancient World, only they still survive.

Their scale is staggering, with a square base averaging 230.4 m (755.8 ft) on an edge. But, as pointed out in Chapter 1 and as summarized in Table 1–2, the accuracy between the lengths of the four sides and the precision of the right angles at each of the four corners of the base is astounding even by modern construction standards and capabilities.

But, there's more. Much more!

Look again at Figures 1–7 and 1–8. Think about the engineering problems that had to be solved. Then think about the techniques that *had* to be employed.

Massive stones (varying in hardness from soft limestone to hard granite and basalt) had to be cut from quarries nearly 13 km (over 8 mi) from the Giza plateau; then each had to be hand hewn to precise shape and dimensions using copper chisels (as iron had not yet been refined, as stone hammers would slow progress and limit precision). Yet no trace of such copper tools has ever been found in the quarries! Notches used to help cleave the great stones from the walls of quarries also aided in the huge blocks being pulled by ropes to the construction site.

Wooden skids and rollers were likely used to reduce the power needed to move the blocks along specially prepared roadways. But, then, the real engineering marvel had to occur—over and over again—2.4 million times to position 2.4 million blocks weighing 2.5 metric tonnes (5500 lb) or more. Once every 2 minutes for 10 hours a day, 365 days per year for 20 years, stones were moved into position on the rising edifice! But how did the ancient Egyptian builders move the great blocks into position? Using human-powered hoists? Up temporary ramps of sand? Or, perhaps, they did not even lift cut stones, as suggested by one recent hypothesis that blocks were cast in place from crushed rock to make "cast stone."[2]

The marvel goes on. While mostly solid, there are burial chambers and inclined access tunnels within the pyramids. Were these cut—or, more likely—were they built in by design and proper layout of the blocks, layer by layer, like a gigantic stereolithographic creation (see Chapter 22)? Once completed, how were the walls of the sacred burial chamber for the beloved pharaoh decorated with elaborate paintings and hieroglyphics depicting the pharaoh's life—and entrance to his new afterlife? If torches were used for lighting, where is the soot? There is none! If torches were not used, how did the ancient artists see to do their work?

We could go on and on reflecting on these timeless engineering masterpieces. But, instead, let's consider one more facet relevant to the purpose of this book before considering a final illustrative example.

The ancient Egyptians employed many of the problem-solving techniques discussed, including mathematical approaches, physical/mechanical approaches, graphic approaches, and abstract approaches. Did they not?

They had to use equations (Chapter 3), including one that stated that the sum of the squares of the legs of a right triangle equals the square of the triangle's hypotenuse, that is, the pythagorean theorem. But they did it 2000 years before Pythagoras of Samos (570–495 BC) was born! Without question, they used interpolation and extrapolation (Chapter 5), similarity and ratios (Chapter 8), scaling (Chapter 10), and other math-based techniques.

---

[2]Michael Barsoum, Distinguished Professor of Materials Science and Engineering at Drexel University, studied a proposed possibility that the limestone blocks used to build at least the upper tiers of the pyramids of Giza were cast, in other words, were a "concrete stone." Interested readers are encouraged to look into some of the startling findings.

There is evidence they used the mechanical approach of reverse engineering (Chapter 16), as the angle of inclination of the walls of the Great Pyramids (at about 52°) evolved from refinements of earlier, smaller pyramids.[3] Inevitably, they used proof-of-concept models (Chapter 18) and test models (Chapter 20), and built earlier prototypes (Chapter 22). Without question, as clever as they were—and they were more clever than most archeologists are inclined to think—they also used some trial and error (Chapter 23), albeit based on guided empiricism, as opposed to blind luck!

It is unimaginable that they did not use many graphic approaches, including sketches and renderings (Chapter 24), layouts (Chapter 27), and, for such a massive, involved, and long-term project, they almost certainly used planning tools such as flowcharts (Chapter 28) and decision trees (Chapter 30), and maybe even storyboards (Chapter 37).

Finally, and most intriguing of all, they used some abstract approaches that may never be unraveled. They had a vision of what needed to be built, of how the three Great Pyramids needed to be aligned with the stars, and how they needed to be oriented relative to one another. The fact that the perimeter of the base of a pyramid is equal to the circumference of a circle whose radius is the altitude of the pyramid is profoundly mystical. Born of some abstract idea, the pyramid had spiritual meaning beyond practical purpose. It is the abstract idea—and meaning—that elevates the Great Pyramids of Giza to the highest level of Maslow's hierarchy of human needs, and may even represent attainment of self-transcendence.

Thomas Jefferson (April 13, 1743–July 4, 1826) wrote of ". . . life, liberty, and the pursuit of happiness." Attainments, such as by design and problem-solving at the lower three levels of Maslow's hierarchy, satisfy the pursuits of life and liberty. However, reaching the highest two levels are the only way to pursue and achieve true happiness.

Illustrative Example CT-1: A Theory for Pyramid Construction

Innumerable theories of how the Great Pyramids of Giza were constructed by the ancient Egyptians have been proposed over the millennia since their construction almost 4600 years ago. Many involved the use of specially constructed external ramps on one or more sides of the structure of sand and/or stone up which the massive quarried stones for the pyramid were dragged by teams of hundreds of men. Other theories proposed the use of scores and scores of various lifting machines that attached to the faces of the rising structure, perhaps using some form of block and tackle (for mechanical advantage), even though these were not thought to have existed at the time. Such a theory was suggested by the Greek Herodotus, known as the "Father of History," in the fifth century BC, who wrote that ". . . the Egyptians transported great stones to the construction site from quarries near the Nile River," floating the stones up the north-flowing river on giant barges, until they had to be dragged over "a specially made causeway, as great a construction as the pyramids" by "100,000 men working in six-month shifts," and then lifting the stones "from step to step" with "machines made of short wooden planks."

The problem, from an engineering standpoint, with theories involving external ramps is that construction of such ramps would involve the movement of as much material as was used in the pyramid itself—twice: once to create the ramps and again to remove them. The problem with theories involving external scaffolding and lifting devices is that such structures would prevent the rising structure from being observed using sighting instruments.

So external ramps don't seem practical, and lifting devices raise questions as to how the architects monitored the structure as it climbed higher and higher, yet still achieved mind-boggling precision. And any self-respecting engineer or engineer-to-be doesn't—or shouldn't—believe for one moment that the Egyptians were able to "levitate" the stones, with or without help from ancient aliens!

---

[3] It also turns out that the angle chosen is close to the angle formed by a freestanding pile of sand!

In 1999, Henri Houdin, a retired civil engineer from France, enlisted the help of his son, Jean-Pierre Houdin, an established architect with expertise in 3D graphics, to develop his idea that the pyramids had been built from the inside out. During 2000, the two met with members of a French team that, in 1986, had worked on the mystery of the Pyramid of Khufu under the aegis of the Foundation EDF. Two startling findings lent credence to Henri's theory. First, Professor Huy Duong Bui had shown the surveying team plans on which were discovered what they first believed was a construction anomaly for which earlier hypotheses of pyramid construction could not account. The anomaly dubbed "the spiral structure," looked exactly like a ramp built inside the pyramid that would have played a part—perhaps a major part—in the pyramid's construction. Second, the team fortuitously spotted a desert fox in a hole next to a 90-m-high notch high up on the pyramid near one edge between faces. Short of scaling the steep face, how else could the fox have gotten there other than by navigating some kind of internal passageway?

In 2003, Henri created the Association of the Construction of the Great Pyramid (ACGP) in order to promote his research. The association enabled Henri and Jean-Pierre to meet a number of experts, and the project—rapidly morphing into an obsession—moved ahead.

In 2005, Mehdi Tayoubi and Richard Breitner of Dassault Systemes invited Jean-Pierre, who took over the project from his father, to join a new sponsorship program called "Passion for Innovation." Together they decided to examine the theory with the aid of Dassault Systemes' industrial and scientific 3D solutions. Using software applications like CATIA to reconstitute the site of the gigantic construction in three dimensions allowed them to test in real time whether such an approach was plausible. The result is the spectacular *Khufu Reborn,* "a 3-D Experience," that has played in one of the greatest virtual reality theaters in the world, La Geode in Paris, and, fortunately for those of us not able to get to Paris, on the Internet at www.3ds.com/khufu.

Readers are strongly encouraged to view this remarkable documentary, updated in 2011 from an earlier 2007 version.

Figures CT-1, CT-2, and CT-3 show three stills from *Khufu Reborn.* The first image (Figure CT-1) shows a rendering of the huge pyramid in the progress of construction. A single external ramp is depicted that allows stones to be pulled up to a certain level, from which point they are dragged up the hypothesized internal ramp to build higher and higher tiers. The internal ramp can barely be seen near the left-hand corner of the huge lower platform. The second image (Figure CT-2) shows a solid model of the internal ramp, as

Figure CT-1

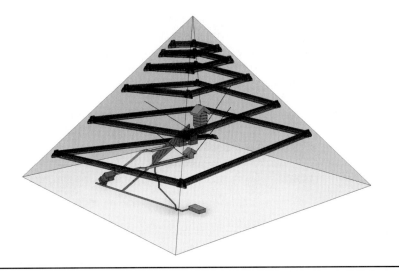

it spirals around the inside of the pyramid, just below the outer facing. The third image (Figure CT-3) depicts the double-level nature of the internal ramps, the lower one to allow teams to drag blocks to allow placement and the upper one to allow teams to return to assist with turning the hue stones at corners—with the aid of lifting devices—and, then, back out of the pyramid to collect another stone from the marshalling area.

Fascinating!

But enough of pyramids . . .

This book was written by an engineer who chose—28 years ago—to teach what he had learned as a practicing engineer for 16 years. More than 50 techniques have been identified, described, discussed, and exemplified. These may—or may not—be all-inclusive. The author thinks they are all-inclusive, having used most, if not all, with success. Which technique to use when? That comes more easily with experience. But the good news is, there is usually "more than one way to skin a cat," to quote a phrase that would upset PETA.

Figure CT-3

Relish being an engineer. Practice the profession as honorably as you can and as well as you are able. The public depends on you. Equip yourself with the "techniques of the profession." Learn—or, at least, familiarize yourself with—the variety of techniques that can be used. A good kit of tools doesn't ensure a skilled craftsperson, but a skilled craftsperson has to have a good kit of tools!

Look forward to becoming an engineer, and enjoy being an engineer! Don't look for fame, but avoid infamy! Do your best—always! Be aware of the techniques that have served your predecessors for decades, centuries, millennia. Good techniques are timeless. The ones presented herein are also time-tested. Pick and choose!

And, last but not least—look yet again at the front cover and ponder all the skills, all the dedication, all the hard work, and all the techniques used to succeed with edifices that still amaze and mystify engineers and place human beings so far ahead of our fellow creatures on this planet.

Robert W. Messler, Jr.

Suggested Resources

Brier, Bob, and Jean-Pierre Houdin, *The Secret of the Great Pyramid: One Man's Obsession Led to the Solution of Ancient Egypt's Greatest Mystery,* HarperCollins, New York, 2008.

Houdin, Jean-Pierre, *Khufu: The Secrets Behind the Building of the Great Pyramid,* Farid Atiya Press, Cairo, Egypt, 2008.

Index